Introduction to Environmental Engineering

Introduction to Environmental Engineering

Introduction to Environmental Engineering

Editor: Grant Paterson

www.callistoreference.com

Callisto Reference,
118-35 Queens Blvd., Suite 400,
Forest Hills, NY 11375, USA

Visit us on the World Wide Web at:
www.callistoreference.com

ISBN: 978-1-64116-014-8 (Hardback)

Cataloging-in-Publication Data

Introduction to environmental engineering / edited by Grant Paterson.
    p. cm.
Includes bibliographical references and index.
ISBN 978-1-64116-014-8
1. Environmental engineering. 2. Sustainable engineering.
3. Environmental protection. I. Paterson, Grant.
TA170 .I58 2018
628--dc23

Table of Contents

Table of Contents

Preface

Environment is essential for our survival and the survival of all the species present on earth. Environmental engineering is the study and process of using the applications of engineering in preserving and protecting the human population and other species by saving the environment. It deals with preventing and minimizing the harmful effects that human activities have on the ecosystem. The topics included in this textbook offer deep insights about the methods and techniques used in this area. Also included herein is a detailed explanation of the various concepts and applications of environmental engineering. It is a complete source of knowledge on the present status of this important field.

A foreword of all Chapters of the book is provided below:

Chapter 1 - Environmental engineering is a part of engineering that focuses on the protection of the environment from adverse man-made effects, as well as humans from the ill effects of natural pollutants. It also lays special emphasis on preserving the environmental quality prone to environmental degradation. This chapter is an overview of the subject matter incorporating all the major aspects of environmental engineering; **Chapter 2 -** Water quality measures the quality of water in context to the need of biotic species related with it. Monitoring the quality of water helps in identifying present and future problems of water pollution, availability of water resources as per usage, developing plans and setting priorities for water quality management program, etc. This section has been carefully written to provide an easy understanding of the varied facets of monitoring water quality; **Chapter 3 -** Biological methods of controlling wastewater production can be divided into aerobic processes and anaerobic processes. Aerobic processes include aerated lagoons, stabilization ponds, aerobic digestion, etc. Anaerobic processes include anaerobic contact process, anaerobic digestion, etc. The major components used for controlling water pollution are discussed in this chapter; **Chapter 4 -** The presence of particles in the air, which can be harmful for living organisms, is referred to as air pollution. It can be classified into natural contaminants, gases and vapors, aerosols, etc. Air pollution can be contained through various methods and concepts such as the use of alternative fuel, combustion control, three-way catalytic convertor, etc. The aspects elucidated in this section are of vital importance, and provide a better understanding of air pollution.

I would like to thank the entire editorial team who made sincere efforts for this book and my family who supported me in my efforts of working on this book. I take this opportunity to thank all those who have been a guiding force throughout my life.

Editor

Environmental Engineering: An Overview

Environmental engineering is a part of engineering that focuses on the protection of the environment from adverse man-made effects, as well as humans from the ill effects of natural pollutants. It also lays special emphasis on preserving the environmental quality prone to environmental degradation. This chapter is an overview of the subject matter incorporating all the major aspects of environmental engineering.

Environmental Engineering

- According to Peavy et al. it is that branch of engineering that is concerned with protecting the environment from the potentially deleterious effects of human activity, protecting human populations from the effects of adverse environmental actors and improving environmental quality for human health and well being.

- Environmental engineering is still an evolving branch of engineering that is closely related to Chemical and Civil engineering.

- It is closely associated with chemistry, physics and biology; and has elements of hydrology, meteorology, atmospheric sciences, environmental chemistry, microbiology and ecology.

Environmental engineering is the branch of engineering concerned with the application of scientific and engineering principles for protection of human populations from the effects of adverse environmental factors; protection of environments, both local and global, from potentially deleterious effects of natural and human activities; and improvement of environmental quality.

Environmental engineering can also be described as a branch of applied science and technology that addresses the issues of energy preservation, production asset and control of waste from human and animal activities. Furthermore, it is concerned with finding plausible solutions in the field of public health, such as waterborne diseases, implementing laws which promote adequate sanitation in urban, rural and recreational areas. It involves waste water management, air pollution control, recycling, waste disposal, radiation protection, industrial hygiene, animal agriculture, environmental sustainability, public health and environmental engineering law. It also includes studies on the environmental impact of proposed construction projects.

Environmental engineers study the effect of technological advances on the environment. To do so, they conduct studies on hazardous-waste management to evaluate the significance of such hazards, advise on treatment and containment, and develop regulations to prevent mishaps. Environmental engineers design municipal water supply and industrial wastewater treatment systems. They address local and worldwide environmental issues such as the effects of acid rain, global warming, ozone depletion, water pollution and air pollution from automobile exhausts and industrial sources.

At many universities environmental engineering programs are offered at either the department of civil engineering or the department of chemical engineering at engineering faculties. Environmental "civil" engineers focus on hydrology, water resources management, bioremediation, and water treatment plant design. Environmental "chemical" engineers, on the other hand, focus on environmental chemistry, advanced air and water treatment technologies and separation processes.

Additionally, engineers are more frequently obtaining specialized training in law (J.D.) and are utilizing their technical expertise in the practices of environmental engineering law.

Most jurisdictions also impose licensing and registration requirements.

Development

Ever since people first recognized that their health is related to the quality of their environment, they have applied principles to attempt to improve the quality of their environment. The ancient Indian Harappan civilization utilized early sewers in some cities more than 5000 years ago. The Romans constructed aqueducts to prevent drought and to create a clean, healthful water supply for the metropolis of Rome. In the 15th century, Bavaria created laws restricting the development and degradation of alpine country that constituted the region's water supply.

The field emerged as a separate environmental discipline during the middle third of the 20th century in response to widespread public concern about water and pollution and increasingly extensive environmental quality degradation. However, its roots extend back to early efforts in public health engineering. Modern environmental engineering began in London in the mid-19th century when Joseph Bazalgette designed the first major sewerage system that reduced the incidence of waterborne diseases such as cholera. The introduction of drinking water treatment and sewage treatment in industrialized countries reduced waterborne diseases from leading causes of death to rarities.

In many cases, as societies grew, actions that were intended to achieve benefits for those societies had longer-term impacts which reduced other environmental qualities. One example is the widespread application of the pesticide DDT to control agricultural pests in the years following World War II. While the agricultural benefits were outstanding and crop yields increased dramatically thus reducing world hunger substantially, and malaria was controlled better than it ever had been, numerous species were brought to the verge of extinction due to the impact of the DDT on their reproductive cycles. The story of DDT as vividly told in Rachel Carson's *Silent Spring* (1962) is considered to be the birth of the modern environmental movement and of the modern field of "environmental engineering."

Conservation movements and laws restricting public actions that would harm the environment have been developed by various societies for millennia. Notable examples are the laws decreeing the construction of sewers in London and Paris in the 19th century and the creation of the U.S. national park system in the early 20th century.

Scope

Solid Waste Management

Environmental Impact Assessment and Mitigation

Scientists have air pollution dispersion models to evaluate the concentration of a pollutant at a receptor or the impact on overall air quality from vehicle exhausts and industrial flue gas stack emissions. To some extent, this field overlaps the desire to decrease carbon dioxide and other greenhouse gas emissions from combustion processes. They apply scientific and engineering principles to evaluate if there are likely to be any adverse impacts to water quality, air quality, habitat quality, flora and fauna, agricultural capacity, traffic impacts, social impacts, ecological impacts, noise impacts, visual (landscape) impacts, etc. If impacts are expected, they then develop mitigation measures to limit or prevent such impacts. An example of a mitigation measure would be the creation of wetlands in a nearby location to mitigate the filling in of wetlands necessary for a road development if it is not possible to reroute the road.

In the United States, the practice of environmental assessment was formally initiated on January 1, 1970, the effective date of the National Environmental Policy Act (NEPA). Since that time, more than 100 developing and developed nations either have planned specific analogous laws or have adopted procedure used elsewhere. NEPA is applicable to all federal agencies in the United States.

Water Supply and Treatment

They evaluate the water balance within a watershed and determine the available water supply, the water needed for various needs in that watershed, the seasonal cycles of water movement through the watershed and they develop systems to store, treat, and convey water for various uses. Water is treated to achieve water quality objectives for the end uses. In the case of a potable water supply, water is treated to minimize the risk of infectious disease transmission, the risk of non-infectious illness, and to create a palatable water flavor. Water distribution systems are designed and built to provide adequate water pressure and flow rates to meet various end-user needs such as domestic use, fire suppression, and irrigation.

Sewage treatment plant, Australia

Wastewater Treatment

Water pollution

There are numerous wastewater treatment technologies. A wastewater treatment train can consist of a primary clarifier system to remove solid and floating materials, a secondary treatment system consisting of an aeration basin followed by flocculation and sedimentation or an activated sludge system and a secondary clarifier, a tertiary biological nitrogen removal system, and a final disinfection process. The aeration basin/activated sludge system removes organic material by growing bacteria (activated sludge). The secondary clarifier removes the activated sludge from the water. The tertiary system, although not always included due to costs, is becoming more prevalent to remove nitrogen and phosphorus and to disinfect the water before discharge to a surface water stream or ocean outfall.

Air Pollution Management

Scientists have developed air pollution dispersion models to evaluate the concentration of a pollutant at a receptor or the impact on overall air quality from vehicle exhausts and industrial flue gas stack emissions. To some extent, this field overlaps the desire to decrease carbon dioxide and other greenhouse gas emissions from combustion processes.

Environmental Protection Agency

The U.S. Environmental Protection Agency (EPA) is one of the many agencies that work with environmental engineers to solve key issues. An important component of EPA's mission is to protect and improve air, water, and overall environmental quality in order to avoid or mitigate the consequences of harmful effects.

Ecological Engineering for Sustainable Agriculture in Arid and Semiarid West African Regions

Ecological engineering offers new alternatives for the management of agricultural systems that are more tailored to the ever-changing social and environmental necessities in these regions. This requires managing the complexity of agrosystems, while striving to mimic the functioning of natural ecosystems of West African drylands and taking advantage of traditional practices and local know-how resulting from a long process of adaptation to environmental constraints.

1. Acting on biodiversity. Biodiversity is essential to the productivity of ecosystems and their temporal stability under the impact of external disturbances. Several ecological processes related to biodiversity may be intensified for the benefit of agrosilvopastoral systems: promoting diversity and soil microorganism activity to benefit plants, associating and utilizing the mutual benefits of plants.

2. Utilizing organic matter and nutrient cycles. The productivity of agrosystems with low chemical input use in dryland regions is primarily based on efficient organic resource management, and in turn on the nutrient and energy flows they induce. It is thus possible to intervene at several levels: enhancing crop-livestock farming integration to preserve natural resources, restoring the biological activity of soils via specific organic inputs, supplying nutrients to plants locally.

3. Enhancing available water use. Water supplies are limited and irregular in dryland areas. Current management of these supplies—which involves capturing rainwater and surface runoff—could be improved in several ways: adapting to erratic rainfall or drought risks by focusing on: (i) the organization of the farm and community (farm plot patterns in association with the random rainfall distribution, etc.), and on (ii) cropping techniques to reduce crop water needs (plant choices, weeding, etc.), preserving water in crop fields by hampering runoff, accounting for the essential role of trees regarding soil and water in drylands.

4. Managing landscapes and associated ecological processes. Ecological crop pest regulation by their natural enemies is one ecosystem service provided by biodiversity.

Better pest management could be considered in association with promoting biodiversity at different scales, e.g. from the plant to the landscape.

Education

Courses aimed at developing graduates with specific skills in environmental systems or environmental technology are becoming more common and fall into broad classes:

- *Mechanical engineering* courses oriented towards designing machines and mechanical systems for environmental use such as water treatment facilities, pumping stations, garbage segregation plants and other mechanical facilities;

- *Environmental engineering or environmental systems* courses oriented towards a civil engineering approach in which structures and the landscape are constructed to blend with or protect the environment;

- *Environmental chemistry, sustainable chemistry* or *environmental chemical engineering* courses oriented towards understanding the effects (good and bad) of chemicals in the environment. Focus on mining processes, pollutants and commonly also cover biochemical processes;

- *Environmental technology* courses oriented towards producing electronic or electrical graduates capable of developing devices and artifacts able to monitor, measure, model and control environmental impact, including monitoring and managing energy generation from renewable sources.

Prominent environmental Engineers

- Robert A. Gearheart
- Paul V. Roberts
- Daniel A. Vallero
- Abel Wolman

Environmental Degradation

Environmental degradation is the deterioration of the environment through depletion of resources such as air, water and soil; the destruction of ecosystems; habitat destruction; the extinction of wildlife; and pollution. It is defined as any change or disturbance to the environment perceived to be deleterious or undesirable. As indicated by the I=PAT equation, environmental impact (I) or degradation is caused by the combination of an already very large and increasing human population (P), continually increasing economic growth or per capita affluence (A), and the application of resource depleting and polluting technology (T).

Eighty-plus years after the abandonment of Wallaroo Mines (Kadina, South Australia), mosses remain the only vegetation at some spots of the site's grounds

Environmental degradation is one of the ten threats officially cautioned by the High-level Panel on Threats, Challenges and Change of the United Nations. The United Nations International Strategy for Disaster Reduction defines environmental degradation as "The reduction of the capacity of the environment to meet social and ecological objectives, and needs". Environmental degradation is of many types. When natural habitats are destroyed or natural resources are depleted, the environment is degraded. Efforts to counteract this problem include environmental protection and environmental resources management.

Water Degradation

One major component of environmental degradation is the depletion of the resource of fresh water on Earth. Approximately only 2.5% of all of the water on Earth is fresh water, with the rest being salt water. 69% of the fresh water is frozen in ice caps located on Antarctica and Greenland, so only 30% of the 2.5% of fresh water is available for consumption. Fresh water is an exceptionally important resource, since life on Earth is ultimately dependent on it. Water transports nutrients and chemicals within the biosphere to all forms of life, sustains both plants and animals, and moulds the surface of the Earth with transportation and deposition of materials.

The current top three uses of fresh water account for 95% of its consumption; approximately 85% is used for irrigation of farmland, golf courses, and parks, 6% is used for domestic purposes such as indoor bathing uses and outdoor garden and lawn use, and 4% is used for industrial purposes such as processing, washing, and cooling in manufacturing centers. It is estimated that one in three people over the entire globe are already facing water shortages, almost one-fifth of the world's population live in areas of physical water scarcity, and almost one quarter of the world's population live in a developing country that lacks the necessary infrastructure to use water from available rivers and aquifers. Water scarcity is an increasing problem due to many foreseen issues in the future, including population growth, increased urbanization, higher standards of living, and climate change.

Climate Change and Temperature

Climate change affects the Earth's water supply in a large number of ways. It is predicted that the mean global temperature will rise in the coming years due to a number of forces affecting the climate, the amount of atmospheric CO_2 will rise, and both of these will influence water resources; evaporation depends strongly on temperature and moisture availability, which can ultimately affect the amount of water available to replenish groundwater supplies.

Transpiration from plants can be affected by a rise in atmospheric CO_2, which can decrease their use of water, but can also raise their use of water from possible increases of leaf area. Temperature increase can decrease the length of the snow season in the winter and increase the intensity of snowmelt in warmer seasons, leading to peak runoff of snowmelt earlier in the season, affecting soil moisture, flood and drought risks, and storage capacities depending on the area.

Warmer winter temperatures cause a decrease in snowpack, which can result in diminished water resources during summer. This is especially important at mid-latitudes and in mountain regions that depend on glacial runoff to replenish their river systems and groundwater supplies, making these areas increasingly vulnerable to water shortages over time; an increase in temperature will initially result in a rapid rise in water melting from glaciers in the summer, followed by a retreat in glaciers and a decrease in the melt and consequently the water supply every year as the size of these glaciers get smaller and smaller.

Thermal expansion of water and increased melting of oceanic glaciers from an increase in temperature gives way to a rise in sea level, which can affect the fresh water supply of coastal areas as well; as river mouths and deltas with higher salinity get pushed further inland, an intrusion of saltwater results in an increase of salinity in reservoirs and aquifers. Sea-level rise may also consequently be caused by a depletion of groundwater, as climate change can affect the hydrologic cycle in a number of ways. Uneven distributions of increased temperatures and increased precipitation around the globe results in water surpluses and deficits, but a global decrease in groundwater suggests a rise in sea level, even after meltwater and thermal expansion were accounted for, which can provide a positive feedback to the problems sea-level rise causes to fresh-water supply.

A rise in air temperature results in a rise in water temperature, which is also very significant in water degradation, as the water would become more susceptible to bacterial growth. An increase in water temperature can also affect ecosystems greatly because of a species' sensitivity to temperature, and also by inducing changes in a body of water's self-purification system from decreased amounts of dissolved oxygen in the water due to rises in temperature.

Climate Change and Precipitation

A rise in global temperatures is also predicted to correlate with an increase in global precipitation, but because of increased runoff, floods, increased rates of soil erosion, and mass movement of land, a decline in water quality is probable, while water will carry more nutrients, it will also carry more contaminants. While most of the attention about climate change is directed towards global warming and greenhouse effect, some of the most severe effects of climate change are likely to be from changes in precipitation, evapotranspiration, runoff, and soil moisture. It is generally expected that, on average, global precipitation will increase, with some areas receiving increases and some decreases.

Climate models show that while some regions should expect an increase in precipitation, such as in the tropics and higher latitudes, other areas are expected to see a decrease, such as in the subtropics; this will ultimately cause a latitudinal variation in water distribution. The areas receiving more precipitation are also expected to receive this increase during their winter and actually become drier during their summer, creating even more of a variation of precipitation distribution. Naturally, the distribution of precipitation across the planet is very uneven, causing constant variations in water availability in respective locations.

Changes in precipitation affect the timing and magnitude of floods and droughts, shift runoff processes, and alter groundwater recharge rates. Vegetation patterns and growth rates will be directly affected by shifts in precipitation amount and distribution, which will in turn affect agriculture as well as natural ecosystems. Decreased precipitation will deprive areas of water, causing water tables to fall and reservoirs and wetlands, rivers, and lakes to empty, and possibly an increase in evaporation and evapotranspiration, depending on the accompanied rise in temperature. Groundwater reserves will be depleted, and the remaining water has a greater chance of being of poor quality from saline or contaminants on the land surface.

Population Growth

The human population on Earth is expanding rapidly which goes hand in hand with the degradation of the environment at large measures. Humanity's appetite for needs is disarranging the environment's natural equilibrium. Production industries are venting smoke and discharging chemicals that are polluting water resources. The smoke that is emitted into the atmosphere holds detrimental gases such as carbon monoxide and sulfur dioxide. The high levels of pollution in the atmosphere form layers that are eventually absorbed into the atmosphere. Organic compounds such as chlorofluorocarbons

(CFC's) have generated an unwanted opening in the ozone layer, which emits higher levels of ultraviolet radiation putting the globe at large threat.

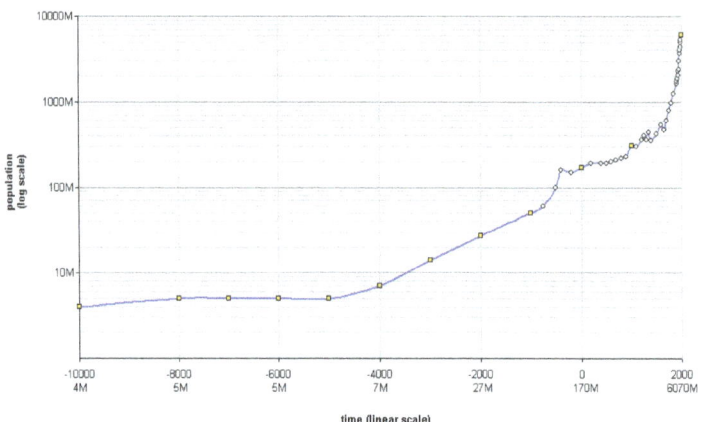

World population growth in a lin-log scale.

The available fresh water being affected by the climate is also being stretched across an ever-increasing global population. It is estimated that almost a quarter of the global population is living in an area that is using more than 20% of their renewable water supply; water use will rise with population while the water supply is also being aggravated by decreases in streamflow and groundwater caused by climate change. Even though some areas may see an increase in freshwater supply from an uneven distribution of precipitation increase, an increased use of water supply is expected.

An increased population means increased withdrawals from the water supply for domestic, agricultural, and industrial uses, the largest of these being agriculture, believed to be the major non-climate driver of environmental change and water deterioration. The next 50 years will likely be the last period of rapid agricultural expansion, but the larger and wealthier population over this time will demand more agriculture.

Population increase over the last two decades, at least in the United States, has also been accompanied by a shift to an increase in urban areas from rural areas, which concentrates the demand for water into certain areas, and puts stress on the fresh water supply from industrial and human contaminants. Urbanization causes overcrowding and increasingly unsanitary living conditions, especially in developing countries, which in turn exposes an increasingly number of people to disease. About 79% of the world's population is in developing countries, which lack access to sanitary water and sewer systems, giving rises to disease and deaths from contaminated water and increased numbers of disease-carrying insects.

Agriculture

Agriculture is dependent on available soil moisture, which is directly affected by climate dynamics, with precipitation being the input in this system and various processes being the output, such as evapotranspiration, surface runoff, drainage, and percolation into groundwater. Changes in climate, especially the changes in precipitation and evapotranspiration predicted by climate models, will directly affect soil moisture, surface runoff, and groundwater recharge.

Water pollution due to dairy farming in the Wairarapa in New Zealand

In areas with decreasing precipitation as predicted by the climate models, soil moisture may be substantially reduced. With this in mind, agriculture in most areas needs irrigation already, which depletes fresh water supplies both by the physical use of the water and the degradation agriculture causes to the water. Irrigation increases salt and nutrient content in areas that would not normally be affected, and damages streams and rivers from damming and removal of water. Fertilizer enters both human and livestock waste streams that eventually enter groundwater, while nitrogen, phosphorus, and other chemicals from fertilizer can acidify both soils and water. Certain agricultural demands may increase more than others with an increasingly wealthier global population, and meat is one commodity expected to double global food demand by 2050, which directly affects the global supply of fresh water. Cows need water to drink, more if the temperature is high and humidity is low, and more if the production system the cow is in is extensive, since finding food takes more effort. Water is needed in processing of the meat, and also in the production of feed for the livestock. Manure can contaminate bodies of freshwater, and slaughterhouses, depending on how well they are managed, contribute waste such as blood, fat, hair, and other bodily contents to supplies of fresh water.

The transfer of water from agricultural to urban and suburban use raises concerns about agricultural sustainability, rural socioeconomic decline, food security, an increased carbon footprint from imported food, and decreased foreign trade balance. The depletion of fresh water, as applied to more specific and populated areas, increases fresh water scarcity among the population and also makes populations susceptible to economic, social, and political conflict in a number of ways; rising sea levels forces migration from coastal areas to other areas farther inland, pushing populations closer together breaching borders and other geographical patterns, and agricultural surpluses and deficits from the availability of water induce trade problems and economies of certain areas. Climate change is an important cause of involuntary migration and forced displacement According to the Food and Agriculture Organization of the United Nations, global greenhouse gas emissions from animal agriculture exceeds that of transportation.

Water Management

A stream in the town of Amlwch, Anglesey which is contaminated by acid mine drainage from the former copper mine at nearby Parys Mountain

The issue of the depletion of fresh water can be met by increased efforts in water management. While water management systems are often flexible, adaptation to new hydrologic conditions may be very costly. Preventative approaches are necessary to avoid high costs of inefficiency and the need for rehabilitation of water supplies, and innovations to decrease overall demand may be important in planning water sustainability.

Water supply systems, as they exist now, were based on the assumptions of the current climate, and built to accommodate existing river flows and flood frequencies. Reservoirs are operated based on past hydrologic records, and irrigation systems on historical temperature, water availability, and crop water requirements; these may not be a reliable guide to the future. Re-examining engineering designs, operations, optimizations, and planning, as well as re-evaluating legal, technical, and economic approaches to manage water resources are very important for the future of water management in response to water degradation. Another approach is water privatization; despite its economic and cultural effects, service quality and overall quality of the water can be more easily controlled and distributed. Rationality and sustainability is appropriate, and requires limits to overexploitation and pollution, and efforts in conservation.

Environmental Ethics and EIA

- Traditionally, industries and its basic components were designed based upon technical and economic considerations only. Now-a-days, it is essential to consider environment, health and safety as factors during design.

- Environmental ethics is related to attitude of people towards other living beings and environment.

- During any project, though it is essential that 'economic sustainability' is attained; however, it is also essential that 'ecological sustainability' and 'social sustainability' are also attained.

- Impact assessment is a handy tool to assess the environmental compatibility of the projects in terms of their location, suitability of technology, efficiency in resources utilization and recycling, etc.

- Environmental Impact Assessment (EIA) has now been made a prerequisite for the settling up of new projects and renewal of licenses of old and existing plants.

- EIA is a major instrument in decision making and for measurement of sustainability in the context of the regional carrying capacity. It provides the conceptual framework for extending the cumulative assessment of development policies, plans and projects on a regional basis.

- Sustainable development of chemical process industries is a process in which the exploitation of resources and the direction of the investments are all made consistent with future as well as present heads.

Pollution Due to Chemical Process Industries

The primary causes of industrial pollution are :

- Use of outdated and inefficient technologies for product manufacturing, pollution abatement and various other operation in industries which generate a large amount of wastes

- Development of unplanned industrial conglomerations without foreseeing the effect on environment

- The existence of large number of small scale industries without defining land use patterns and environmental regulations for them

- Poor enforcement of pollution control laws for big and small industries.

Major polluting industrial sectors		
1) Cement	2) Thermal power plants	3) Iron & Steel
4) Fertilizer	5) Zinc Smelters	6) Copper Smelters
7) Aluminum Smelters	8) Oil Refineries	9) Distilleries
10) Pulp & Paper	11) Dyes and Dye Intermediates	12) Pesticides
13) Petro Chemicals	14) Petroleum refining	15) Sugar
16) Tanneries	17) Basic Drugs	

Major Concerns of Industrial Pollution

- Water and air pollution from chemical process industries need immediate attention.

- Industrial wastewaters vary widely in their composition and treatment methods, which have to take in to consideration the specific characteristic of the wastes.

- Many treatment practices have followed the approach of mixing the liquid sewage waste with industrial waste and treating the mixture by conventional methods.

- Treatment methods such as lagoon (aerobic & anaerobic), oxidation ditches and aerated lagoons have also been tried with varying degree of success. The majority of treatment plants have, however, failed to succeed. The chief reasons for this have been the omission of some of the key parameters that govern biological oxidation when industrial wastes are treated.

- Physico-chemical methods are necessary to remove or recover the chemical ingredients present in liquid effluents discharged from electroplating, chlor-alkali, pesticides, fertilizers, dyes and pigments, metallurgical, paper and pulp, etc. and other such process industries.

- The reuse of water in processes where the water quality standards are not stringent is worth considering. A considerable quantity of water is presently being reused in process industries in India but a lot more needs to be done in this area.

Major Definitions as Per Indian Environmental Acts

- "Environment" includes water, air and land and the inter-relationship which exists among and between water, air and land and human beings, other living creatures, plants, micro-organism and property.

- "Environmental pollutant" means any solid, liquid or gaseous substance present in such concentration and may be, or tend to be, injurious to environment.

- "Air pollutants" means any solid, liquid or gaseous substance (including noise) present in the atmosphere in such concentration as may be or tend to be injurious to human being or other living creatures or plants or property or environment.

- "Air pollution" means the presence in the atmosphere of any air pollutant

- "Ambient air" means that portion of the atmosphere, external to buildings, to which the general public has access.

Environmental Crisis Due to Industrial Development

- Large scale contamination of water and air.

- Deforestation

- Increase in urban slums

- Generation of huge solid waste consisting of hazardous material.

- Water scarcity and ground water depletion.

- Global warming

- Greenhouse effect

- Ozone layer depletion

Environmental Engineering Science

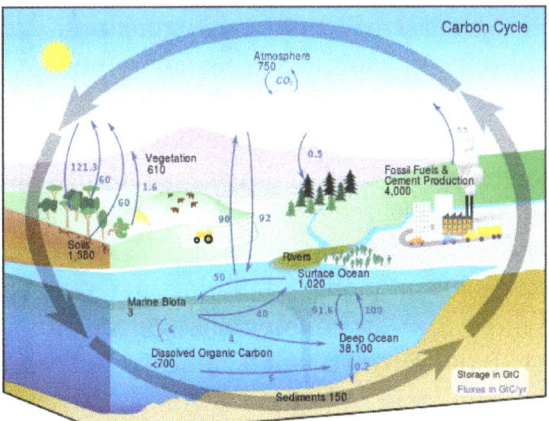

Students in Environmental Engineering Science typically combine scientific studies of the biosphere with mathematical, analytical and design tools found in the engineering fields

Environmental engineering science (EES) is a multidisciplinary field of engineering science that combines the biological, chemical and physical sciences with the field of engineering. This major traditionally requires the student to take basic engineering classes in fields such as thermodynamics, advanced math, computer modeling and simulation and technical classes in subjects such as statics, mechanics, hydrology, and fluid dynamics. As the student progresses, the upper division elective classes define a specific field of study for the student with a choice in a range of science, technology and engineering related classes.

Difference with Related Fields

Graduates of Environmental Engineering Science can go on to work on the technical aspects of designing a Living Roof like the one pictured here at the California Academy of the Sciences

As a recently created program, environmental engineering science has not yet been incorporated into the terminology found among environmentally focused professionals. In the few engineering colleges that offer this major, the curriculum shares more classes in common with environmental engineering than it does with environmental science. Typically, EES students follow a similar course curriculum with environmental engineers until their fields diverge during the last year of college. The majority of the environmental engineering students must take classes designed to connect their knowledge of the environment to modern building materials and construction methods. This is meant to direct the environmental engineer into a field where they will more than likely assist in building treatment facilities, preparing environmental impact assessments or helping to mitigate air pollution from specific point sources.

Meanwhile, the environmental engineering science student will choose a direction for their career. From the range of electives they have to choose from, these students can move into a fields such as the design of nuclear storage facilities, bacterial bioreactors or environmental policies. These students combine the practical design background of an engineer with the detailed theory found in many of the biological and physical sciences.

Description at Universities

Stanford University

The Civil and Environmental Engineering department at Stanford University provides the following description for their program in Environmental Engineering and Science: The Environmental Engineering and Science (EES) program focuses on the chemical and biological processes involved in water quality engineering, water and air pollution, remediation and hazardous substance control, human exposure to pollutants, environmental biotechnology, and environmental protection.

UC Berkeley

The College of Engineering at UC Berkeley defines Environmental Engineering Science, including the following:

This is a multidisciplinary field requiring an integration of physical, chemical and biological principles with engineering analysis for environmental protection and restoration. The program incorporates courses from many departments on campus to create a discipline that is rigorously based in science and engineering, while addressing a wide variety of environmental issues. Although an environmental engineering option exists within the civil engineering major, the engineering science curriculum provides a more broadly based foundation in the sciences than is possible in civil engineering

Massachusetts Institute of Technology

Wet labs are required as part of the lower division curriculum

At MIT, the major is described in their curriculum, including the following:

The Bachelor of Science in Environmental Engineering Science emphasizes the fundamental physical, chemical, and biological processes necessary for understanding the interactions between man and the environment. Issues considered include the provision of clean and reliable water supplies, flood forecasting and protection, development of renewable and nonrenewable energy sources, causes and implications of climate change, and the impact of human activities on natural cycles

Lower Division Coursework

Lower division coursework in this field requires the student to take several laboratory-based classes in calculus-based physics, chemistry, biology, programming and analysis. This is intended to give the student background information in order to introduce them to the engineering fields and to prepare them for more technical information in their upper division coursework.

Upper Division Coursework

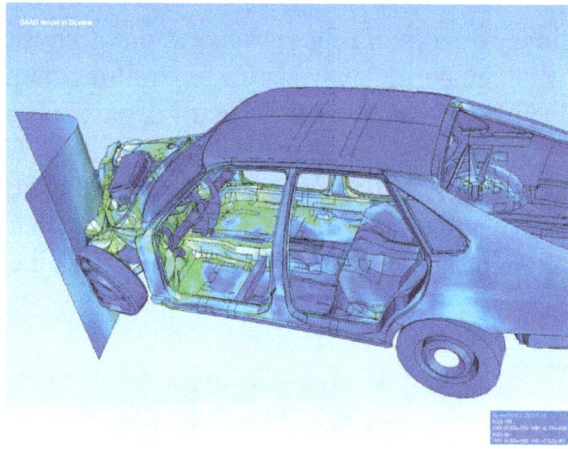

Students learn to integrate their math and statistics with software to perform analysis of physical systems like the Finite Element Analysis shown above

The upper division classes in Environmental Engineering Science prepares the student for work in the fields of engineering and science with coursework in subjects including the following:

- Fluid mechanics

- Mechanics of materials

- Thermodynamics

- Environmental engineering

- Advanced math and statistics

- Geology

- Physical, organic and atmospheric chemistry

- Biochemistry

- Microbiology

- Ecology

Electives

Process Engineering

On this track, students are introduced to the fundamental reaction mechanisms in the field of chemical and biochemical engineering.

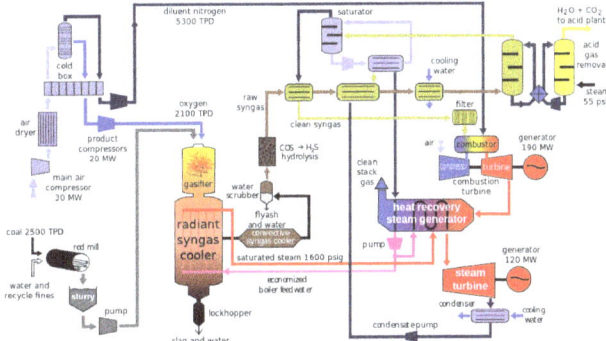

Considering a more environmentally friendly process for coal gasification

Resource Engineering

For this track, students take classes introducing them to ways to conserve natural resources. This can include classes in water chemistry, sanitation, combustion, air pollution and radioactive waste management.

Using design knowledge to make better wastewater treatment facilities

Designing a safe way to store nuclear waste

Geoengineering

This examines geoengineering in detail.

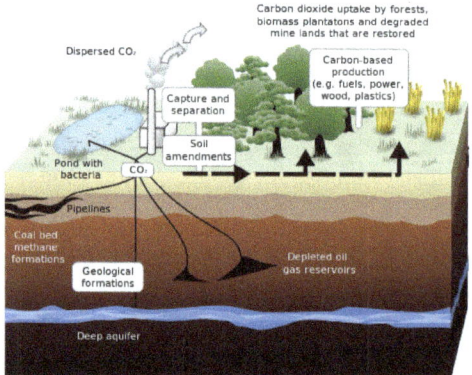

Sequestering carbon from the atmosphere

Ecology

This prepares the students for using their engineering and scientific knowledge to solve the interactions between plants, animals and the biosphere.

How to alter certain biological interactions in order to optimise survival of the system

Examining how the harvesting of kelp effects the sea otter population

Biology

This includes further education about microbial, molecular and cell biology. Classes can include cell biology, virology, microbial and plant biology

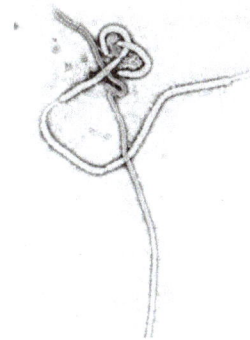

Understanding the way in which viruses function in order to safely sanitize water supplies

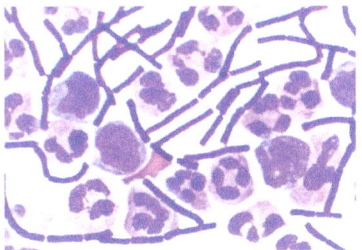

Understanding the metabolism of bacteria in order to see how their proliferation effects the climate

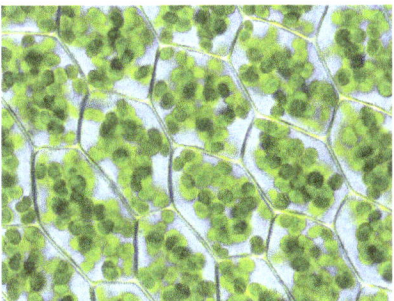

Using the biological design of chloroplasts to design a more effective way of turning solar energy into future sources of power

Policy

This covers in more detail ways the environment can be protected through political means. This is done by introducing students to qualitative and quantitative tools in classes such as economics, sociology, political science and energy and resources.

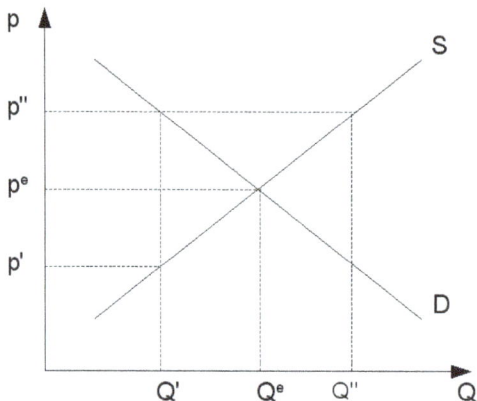

Learning about economics to determine the financial burden it might take to implement an "environmentally friendly" technology

Post Graduation Work

The multidisciplinary approach in Environmental Engineering Science gives the student expertise in technical fields related to their own personal interest. While some graduates choose to use this major to go to graduate school, students who choose to work often go into the fields of civil and environmental engineering, biotechnology, and research. However, the less technical math, programming and writing background gives the students opportunities to pursue IT work and technical writing.

References

- Tchobanoglous, G.; Burton, F.L. & Stensel, H.D. (2003). Wastewater Engineering (Treatment Disposal Reuse) / Metcalf & Eddy, Inc. (4th ed.). McGraw-Hill Book Company. ISBN 0-07-041878-0

- Johnson, D.L., S.H. Ambrose, T.J. Bassett, M.L. Bowen, D.E. Crummey, J.S. Isaacson, D.N. Johnson, P. Lamb, M. Saul, and A.E. Winter-Nelson. 1997. Meanings of environmental terms. Journal of Environmental Quality 26: 581–589

- Turner, D.B. (1994). Workbook of atmospheric dispersion estimates: an introduction to dispersion modeling (2nd ed.). CRC Press. ISBN 1-56670-023-X

- "Architecture and Engineering Occupations : Occupational Outlook Handbook : U.S. Bureau of Labor Statistics". Bls.gov. 2012-03-29. Retrieved 2013-07-01

- Davis, M. L. and D. A. Cornwell, (2006) Introduction to environmental engineering (4th ed.) McGraw-Hill ISBN 978-0072424119

- Powell, Fannetta. "Environmental Degradation and Human Disease". Lecture. SlideBoom. 2009. Web. Retrieved 2011-11-14

2

Water Quality: Monitoring and Treatment

Water quality measures the quality of water in context to the need of biotic species related with it. Monitoring the quality of water helps in identifying present and future problems of water pollution, availability of water resources as per usage, developing plans and setting priorities for water quality management program, etc. This section has been carefully written to provide an easy understanding of the varied facets of monitoring water quality.

Water Quality

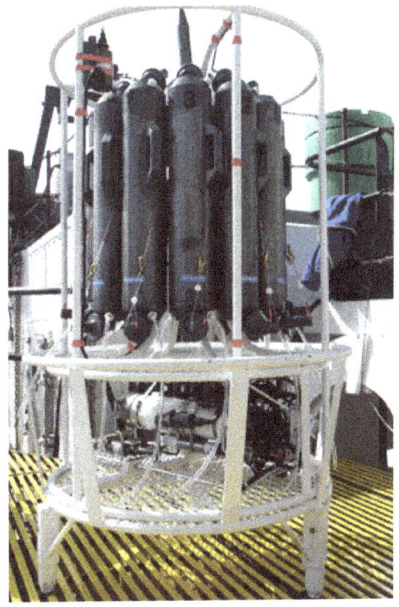

A rosette sampler is used to collect water samples in deep water, such as the Great Lakes or oceans, for water quality testing

Water quality refers to the chemical, physical, biological, and radiological characteristics of water. It is a measure of the condition of water relative to the requirements of one or more biotic species and or to any human need or purpose. It is most frequently used by reference to a set of standards against which compliance can be assessed. The most common standards used to assess water quality relate to health of ecosystems, safety of human contact, and drinking water.

Standards

In the setting of standards, agencies make political and technical/scientific decisions about how the water will be used. In the case of natural water bodies, they also make some reasonable estimate of pristine conditions. Natural water bodies will vary in response to environmental conditions. Environmental scientists work to understand how these systems function, which in turn helps to identify the sources and fates of contaminants. Environmental lawyers and policymakers work to define legislation with the intention that water is maintained at an appropriate quality for its identified use.

The vast majority of surface water on the Earth is neither potable nor toxic. This remains true when seawater in the oceans (which is too salty to drink) is not counted. Another general perception of water quality is that of a simple property that tells whether water is polluted or not. In fact, water quality is a complex subject, in part because water is a complex medium intrinsically tied to the ecology of the Earth. Industrial and commercial activities (e.g. manufacturing, mining, construction, transport) are a major cause of water pollution as are runoff from agricultural areas, urban runoff and discharge of treated and untreated sewage.

Categories

The parameters for water quality are determined by the intended use. Work in the area of water quality tends to be focused on water that is treated for human consumption, industrial use, or in the environment.

Human Consumption

Contaminants that may be in untreated water include microorganisms such as viruses, protozoa and bacteria; inorganic contaminants such as salts and metals; organic chemical contaminants from industrial processes and petroleum use; pesticides and herbicides; and radioactive contaminants. Water quality depends on the local geology and ecosystem, as well as human uses such as sewage dispersion, industrial pollution, use of water bodies as a heat sink, and overuse (which may lower the level of the water).

The United States Environmental Protection Agency (EPA) limits the amounts of certain contaminants in tap water provided by US public water systems. The Safe Drinking Water Act authorizes EPA to issue two types of standards:

- *primary standards* regulate substances that potentially affect human health;

- *secondary standards* prescribe aesthetic qualities, those that affect taste, odor, or appearance.

The U.S. Food and Drug Administration (FDA) regulations establish limits for contaminants in bottled water that must provide the same protection for public health.

Drinking water, including bottled water, may reasonably be expected to contain at least small amounts of some contaminants. The presence of these contaminants does not necessarily indicate that the water poses a health risk.

In urbanized areas around the world, water purification technology is used in municipal water systems to remove contaminants from the source water (surface water or groundwater) before it is distributed to homes, businesses, schools and other recipients. Water drawn directly from a stream, lake, or aquifer and that has no treatment will be of uncertain quality.

Industrial and Domestic use

Dissolved minerals may affect suitability of water for a range of industrial and domestic purposes. The most familiar of these is probably the presence of ions of calcium (Ca^{2+}) and magnesium (Mg^{2+}) which interfere with the cleaning action of soap, and can form hard sulfate and soft carbonate deposits in water heaters or boilers. Hard water may be softened to remove these ions. The softening process often substitutes sodium cations. Hard water may be preferable to soft water for human consumption, since health problems have been associated with excess sodium and with calcium and magnesium deficiencies. Softening decreases nutrition and may increase cleaning effectiveness. Various industries' wastes and effluents can also pollute the water quality in receiving bodies of water.

Environmental Water Quality

Urban runoff discharging to coastal waters

Environmental water quality, also called ambient water quality, relates to water bodies such as lakes, rivers, and oceans. Water quality standards for surface waters vary significantly due to different environmental conditions, ecosystems, and intended human uses. Toxic substances and high populations of certain microorganisms can present a health hazard for non-drinking purposes such as irrigation, swimming, fishing, rafting, boating, and industrial uses. These conditions may also affect wildlife, which use the water for drinking or as a habitat. Modern water quality laws generally specify protection of fisheries and recreational use and require, as a minimum, retention of current quality standards.

Satirical cartoon by William Heath, showing a woman observing monsters in a drop of London water (at the time of the *Commission on the London Water Supply* report, 1828).

There is some desire among the public to return water bodies to pristine, or pre-industrial conditions. Most current environmental laws focus on the designation of particular uses of a water body. In some countries these designations allow for some water contamination as long as the particular type of contamination is not harmful to the designated uses. Given the landscape changes (e.g., land development, urbanization, clearcutting in forested areas) in the watersheds of many freshwater bodies, returning to pristine conditions would be a significant challenge. In these cases, environmental scientists focus on achieving goals for maintaining healthy ecosystems and may concentrate on the protection of populations of endangered species and protecting human health.

Sampling and Measurement

The complexity of water quality as a subject is reflected in the many types of measurements of water quality indicators. The most accurate measurements of water quality are made on-site, because water exists in equilibrium with its surroundings. Measurements commonly made on-site and in direct contact with the water source in question include temperature, pH, dissolved oxygen, conductivity, oxygen reduction potential (ORP), turbidity, and Secchi disk depth.

Sample Collection

An automated sampling station installed along the East Branch Milwaukee River, New Fane, Wisconsin. The cover of the 24-bottle autosampler (center) is partially raised, showing the sample bottles inside. The autosampler was programmed to collect samples at time intervals, or proportionate to flow over a specified period. The data logger (white cabinet) recorded temperature, specific conductance, and dissolved oxygen levels.

More complex measurements are often made in a laboratory requiring a water sample to be collected, preserved, transported, and analyzed at another location. The process of water sampling introduces two significant problems:

- The first problem is the extent to which the sample may be representative of the water source of interest. Many water sources vary with time and with location. The measurement of interest may vary seasonally or from day to night or in response to some activity of man or natural populations of aquatic plants and animals. The measurement of interest may vary with distances from the water boundary with overlying atmosphere and underlying or confining soil. The sampler must determine if a single time and location meets the needs of the investigation, or if the water use of interest can be satisfactorily assessed by averaged values with time and/or location, or if critical maxima and minima require individual measurements over a range of times, locations and/or events. The sample collection procedure must assure correct weighting of individual sampling times and locations where averaging is appropriate. Where critical maximum or minimum values exist, statistical methods must be applied to observed variation to determine an adequate number of samples to assess probability of exceeding those critical values.

- The second problem occurs as the sample is removed from the water source and begins to establish chemical equilibrium with its new surroundings – the sample container. Sample containers must be made of materials with minimal reactivity with substances to be measured; and pre-cleaning of sample containers is important. The water sample may dissolve part of the sample container and any residue on that container, or chemicals dissolved in the water sample may sorb onto the sample container and remain there when the water is poured out for analysis. Similar physical and chemical interactions may take place with any pumps, piping, or intermediate devices used to transfer the water sample into the sample container. Water collected from depths below the surface will normally be held at the reduced pressure of the atmosphere; so gas dissolved in the water may escape into unfilled space at the top of the container. Atmospheric gas present in that air space may also dissolve into the water sample. Other chemical reaction equilibria may change if the water sample changes temperature. Finely divided solid particles formerly suspended by water turbulence may settle to the bottom of the sample container, or a solid phase may form from biological growth or chemical precipitation. Microorganisms within the water sample may biochemically alter concentrations of oxygen, carbon dioxide, and organic compounds. Changing carbon dioxide concentrations may alter pH and change solubility of chemicals of interest. These problems are of special concern during measurement of chemicals assumed to be significant at very low concentrations.

Filtering a manually collected water sample (grab sample) for analysis

Sample preservation may partially resolve the second problem. A common procedure is keeping samples cold to slow the rate of chemical reactions and phase change, and analyzing the sample as soon as possible; but this merely minimizes the changes rather than preventing them. A useful procedure for determining influence of sample containers during delay between sample collection and analysis involves preparation for two artificial samples in advance of the sampling event. One sample container is filled with water known from previous analysis to contain no detectable amount of the chemical of interest. This sample, called a "blank", is opened for exposure to the atmosphere when the sample of interest is collected, then resealed and transported to the laboratory with the sample for analysis to determine if sample holding procedures introduced any measurable amount of the chemical of interest. The second artificial sample is collected with the sample of interest, but then "spiked" with a measured additional amount of the chemical of interest at the time of collection. The blank and spiked samples are carried with the sample of interest and analyzed by the same methods at the same times to determine any changes indicating gains or losses during the elapsed time between collection and analysis.

Testing in Response to Natural Disasters and other Emergencies

Inevitably after events such as earthquakes and tsunamis, there is an immediate response by the aid agencies as relief operations get underway to try and restore basic infrastructure and provide the basic fundamental items that are necessary for survival and subsequent recovery. Access to clean drinking water and adequate sanitation is a priority at times like this. The threat of disease increases hugely due to the large numbers of people living close together, often in squalid conditions, and without proper sanitation.

After a natural disaster, as far as water quality testing is concerned there are widespread views on the best course of action to take and a variety of methods can be employed. The key basic water quality parameters that need to be addressed in an emergency are bacteriological indicators of fecal contamination, free chlorine residual, pH, turbidity and possibly conductivity/total dissolved solids. There are a number of portable water test kits on the market widely used by aid and relief agencies for carrying out such testing.

After major natural disasters, a considerable length of time might pass before water quality returns to pre-disaster levels. For example, following the 2004 Indian Ocean tsunami the Colombo-based International Water Management Institute (IWMI) monitored the effects of saltwater and concluded that the wells recovered to pre-tsunami drinking water quality one and a half years after the event. IWMI developed protocols for cleaning wells contaminated by saltwater; these were subsequently officially endorsed by the World Health Organization as part of its series of Emergency Guidelines.

Chemical Analysis

A gas chromatograph-mass spectrometer measures pesticides and other organic pollutants

The simplest methods of chemical analysis are those measuring chemical elements without respect to their form. Elemental analysis for oxygen, as an example, would indicate a concentration of 890,000 milligrams per litre (mg/L) of water sample because water is made of oxygen. The method selected to measure dissolved oxygen should differentiate between diatomic oxygen and oxygen combined with other elements. The comparative simplicity of elemental analysis has produced a large amount of sample data and water quality criteria for elements sometimes identified as heavy metals. Water analysis for heavy metals must consider soil particles suspended in the water sample. These suspended soil particles may contain measurable amounts of metal. Although the particles are not dissolved in the water, they may be consumed by people drinking the water. Adding acid to a water sample to prevent loss of dissolved metals onto the sample container may dissolve more metals from suspended soil particles. Filtration of soil particles from the water sample before acid addition, however, may cause loss of dissolved metals onto the filter. The complexities of differentiating similar organic molecules are even more challenging.

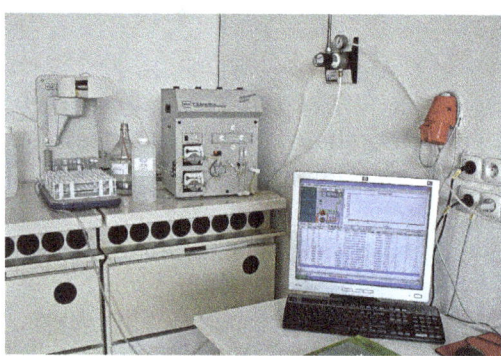

Atomic fluorescence spectroscopy is used to measure mercury and other heavy metals

Making these complex measurements can be expensive. Because direct measurements of water quality can be expensive, ongoing monitoring programs are typically conducted by government agencies. However, there are local volunteer programs and resources available for some general assessment. Tools available to the general public include on-site test kits, commonly used for home fish tanks, and biological assessment procedures.

Real-time Monitoring

Although water quality is usually sampled and analyzed at laboratories, nowadays, citizens demand real-time information about the water they are drinking. During the last years, several companies are deploying worldwide real-time remote monitoring systems for measuring water pH, turbidity or dissolved oxygen levels.

Drinking Water Indicators

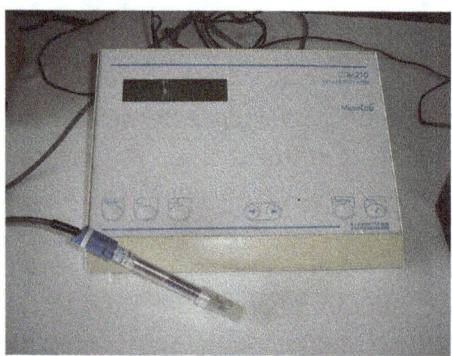

An electrical conductivity meter is used to measure total dissolved solids

The following is a list of indicators often measured by situational category:

- Alkalinity

- Color of water

- pH

- Taste and odor (geosmin, 2-Methylisoborneol (MIB), etc.)

- Dissolved metals and salts (sodium, chloride, potassium, calcium, manganese, magnesium)

- Microorganisms such as fecal coliform bacteria (*Escherichia coli*), Cryptosporidium, and Giardia lamblia

- Dissolved metals and metalloids (lead, mercury, arsenic, etc.)

- Dissolved organics: colored dissolved organic matter (CDOM), dissolved organic carbon (DOC)

- Radon

- Heavy metals

- Pharmaceuticals

- Hormone analogs

Environmental Indicators

Physical Indicators

• Water temperature	• Total dissolved solids (TDS)
• Specific conductance or electrical conductance (EC) or conductivity	• Odour of water
• Total suspended solids (TSS)	• Color of water
• Transparency or turbidity	• Taste of water

Chemical Indicators

• pH	• Heavy metals
• Biochemical oxygen demand (BOD)	• Nitrate
• Chemical oxygen demand (COD)	• Orthophosphates
• Dissolved oxygen (DO)	• Pesticides
• Total hardness (TH)	• Surfactants

Biological Indicators

• Ephemeroptera	• *Escherichia coli* (E. coli)
• Plecoptera	• Coliform bacteria
• Mollusca	
• Trichoptera	

Biological monitoring metrics have been developed in many places, and one widely used measure is the presence and abundance of members of the insect orders Ephemeroptera, Plecoptera and Trichoptera (common names are, respectively, mayfly, stonefly and caddisfly). EPT indexes will naturally vary from region to region, but generally, within a region, the greater the number of taxa from these orders, the better the water quality. Organisations in the United States, such as EPA offer guidance on developing a monitoring program and identifying members of these and other aquatic insect orders.

Individuals interested in monitoring water quality who cannot afford or manage lab scale analysis can also use biological indicators to get a general reading of water quality. One example is the IOWATER volunteer water monitoring program, which includes a benthic macroinvertebrate indicator key.

Bivalve molluscs are largely used as bioindicators to monitor the health of aquatic environments in both fresh water and the marine environments. Their population status or structure, physiology, behaviour or the level of contamination with elements or compounds can indicate the state of contamination status of the ecosystem. They are particularly useful since they are sessile so that they are representative of the environment where they are sampled or placed. A typical project is the Mussel Watch Programme, but today they are used worldwide.

The Southern African Scoring System (SASS) method is a biological water quality monitoring system based on the presence of benthic macroinvertebrates. The SASS aquatic biomonitoring tool has been refined over the past 30 years and is now on the fifth version (SASS5) which has been specifically modified in accordance with international standards, namely the ISO/IEC 17025 protocol. The SASS5 method is used by the South African Department of Water Affairs as a standard method for River Health Assessment, which feeds the national River Health Programme and the national Rivers Database.

Standards and Reports

International

- The World Health Organisation (WHO) has published guidelines for drinking-water quality (GDWQ) in 2011.

- The International Organization for Standardization (ISO) published regulation of water quality in the section of ICS 13.060, ranging from water sampling, drinking water, industrial class water, sewage, and examination of water for chemical, physical or biological properties. ICS 91.140.60 covers the standards of water supply systems.

National Specifications for ambient Water and Drinking Water

European Union

The water policy of the European Union is primarily codified in three directives:

- Directive on Urban Waste Water Treatment (91/271/EEC) of 21 May 1991 concerning discharges of municipal and some industrial wastewaters;

- The Drinking Water Directive (98/83/EC) of 3 November 1998 concerning potable water quality;

- Water Framework Directive (2000/60/EC) of 23 October 2000 concerning water resources management.

India

- Indian Council of Medical Research (ICMR) Standards for Drinking Water.

South Africa

Water quality guidelines for South Africa are grouped according to potential user types (e.g. domestic, industrial) in the 1996 Water Quality Guidelines. Drinking water quality is subject to the South African National Standard (SANS) 241 Drinking Water Specification.

United Kingdom

In England and Wales acceptable levels for drinking water supply are listed in the "Water Supply (Water Quality) Regulations 2000."

United States

In the United States, Water Quality Standards are defined by state agencies for various water bodies, guided by the desired uses for the water body (e.g., fish habitat, drinking water supply, recreational use). The Clean Water Act (CWA) requires each governing jurisdiction (states, territories, and covered tribal entities) to submit a set of biennial reports on the quality of water in their area. These reports are known as the 303(d) and 305(b) reports, named for their respective CWA provisions, and are submitted to, and approved by, EPA. These reports are completed by the governing jurisdiction, typically a state environmental agency. EPA recommends that each state submit a single "Integrated Report" comprising its list of impaired waters and the status of all water bodies in the state. The *National Water Quality Inventory Report to Congress* is a general report on water quality, providing overall information about the number of miles of streams and rivers and their aggregate condition. The CWA requires states to adopt standards for each of the possible designated uses that they assign to their waters. Should evidence suggest or document that a stream, river or lake has failed to meet the water quality criteria for one or more of its designated uses, it is placed on a list of impaired waters. Once a state has placed a water body on this list, it must develop a management plan establishing Total Maximum Daily Loads (TMDLs) for the pollutant(s) impairing the use of the water. These TMDLs establish the reductions needed to fully support the designated uses.

Drinking water standards, which are applicable to public water systems, are issued by EPA under the Safe Drinking Water Act.

Water Quality Monitoring

It is essential for devising water quality management programme to properly use water in any project. It gives information for following decisions to be taken :

- Helps in identifying the present and future problems of water pollution.

- Identifying the present resources of water as per various usages.

- It helps in developing plans and setting priorities for water quality management programme so as to meet future water requirements.

- It helps in evaluating the effectiveness of present management actions being taken and devising future course of actions.

Collection of Water Samples

For physical examination, water can be collected in fully cleaned ordinary buckets or plastic cans. If the water is to be collected for chemicals tests, the container, usually glass bottles of more than 2 liter capacity should be thoroughly washed and cleaned; and then the water should be collected in it.

For the collection of water for bacteriological tests, the person who collects the water must be free from any disease. The containers and bottles must be cleaned with sulphuric acid, potassium dichromate or alkaline permanganate, and then, they should be thoroughly rinsed with distilled water and finally sterilization should be done. Immediately after collection of the samples, bottles should be closed and covered with clot to prevent accumulation of dirt, etc. The testing of water samples should be done as early as possible.

Following points should be kept in view while collecting the samples:

(i) If the water is to be collected from a tap or faucet, sufficient quantity of wastewater should be allowed to pass through the tap, before collecting sample from because it will eliminate the stagnant water.

(ii) If the water is to be collected from the surface stream or river, it should be collected about 40-50 cm below the surface to avoid the collection of surface impurities oils, tree leaves, etc. which should also removed by strainers while collecting the water through intakes.

(iii) In case the water is being collected from the ground sources i.e. through well or tube well, sufficient quantity of water should be pumped out before collecting the samples.

Table: Principal constituents of concern in wastewater treatment [2, 3].

Constituent	Importance
Suspended solids	Lead to sludge deposits and development of anaerobic conditions
Biodegradable organics	Depletion of natural oxygen and to the development of septic condition; Composed principally of proteins, carbohydrates, fats, biodegradable organics, etc.; Measured in terms of biochemical oxygen demand (BOD) and chemical oxygen demand (COD).
Pathogens	Communicable diseases
Nutrients	Nitrogen and phosphorus are principal limiting nutrients for growth; Cause eutrophication in lakes & ponds.
Heavy metals	Added wastewater from commercial and industrial activities; Many of the metals are highly toxic at small concentration also.
Priority pollutants	Organic and inorganic compounds having known or suspected carcinogenicity, mutagenicity, teratogenicity and/or high acute toxicity.
Refractory organics	Organic compounds like surfactants, phenols and agricultural pesticides, etc. resist conventional method of wastewater treatment.
Dissolved inorganics	Inorganic constituents such as calcium, sodium and sulphates are added to the original domestic water supply as a result of water use and may have to be removed if the wastewater is to be reused.

Physical Parameters

The physical tests include the following tests:

Temperature: The temperature of water is measured by means of ordinary thermometers. Density, viscosity, vapor pressure and surface tension of water are all dependent upon the temperature. The saturation values of solids and gases that can be dissolved in water and the rates of chemical, biochemical and biological activity are also determined on the basis of temperature.

The temperature of surface water is generally same as the atmospheric temperature while that of ground water may be more or less than atmospheric temperature.

Color: The color of water is usually due to presence of organic matter in colloid condition, and due to the presence of mineral and dissolved organic and inorganic impurities. Transparent water with a low accumulation of dissolved materials appears blue. Dissolved organic matter such as humus, peat or decaying plant matter, etc. produce a yellow or brown color. Some algae or dinoflagellates produce reddish or deep yellow waters. Water rich in phytoplankton and other algae usually appears green. Soil runoff water has a variety of yellow, red, brown and gray colors.

The color in water is not harmful but it is objectionable. The color of a water sample can be reported as Apparent or True color. Apparent color is the color of the whole water sample and consists of color from both dissolved and suspended components. True color is measured after filtering the water sample to remove all suspended material.

Before testing the color of the water, first of all total suspended matter should removed from the water by centrifugal force in a special apparatus. After this, the color the water is compared with standard color solution or color discs. When multicolored industrial wastes are involved, such color measurement is meaningless.

The color produced by one milligram of platinum in a litre of distilled water has been fixed as the unit of color.

Turbidity: It is caused due to presence of suspended and colloidal matter in the water. Ground waters are generally less turbid than the surface water. The character and amount of turbidity depends on the type of soil over which the water has moved.

Turbidity is a measure of the resistance of water to the passage of light through it. Turbidity is expressed in parts per million (ppm or milligrams per litre or mg/1). Earlier, the turbidity produced by one milligram of silica in one litre of distilled water was considered as the unit of turbidity.

Turbidity was previously determined by Jackson candle Turbidity units (JTU). This unit is now replaced by more appropriate unit called Nephelometric Turbidity unit (NTU) which is the turbidity produced by one milligram of formazin polymer in one litre of distilled water.

Nephelometry method has better sensitivity, precision and applicability over a wide range of particle size and concentrations as compared to older methods.

Tastes and odors: Tastes and odors in water are due to the presence of (i) dead or living micro- organisms; (ii) dissolved gases such as hydrogen sulphide, methane, carbon dioxide or oxygen combined with organic matter; (iii) mineral substances such as sodium chloride, iron compounds; and (iv) carbonates and sulphates.

The odor of water also changes with temperature. The odor may be classified as sweetish, vegetable, greasy, etc. The odor of both cold and hot water should be determined.

The intensities of the odors are measured in terms of threshold odor number (TON).

TON indicates how many dilutions it takes to produce odor-free water. In this method, enough odor- free water is added to the flasks containing different amount of sample to create a total volume of 200 mL.

$$TON = \frac{A+B}{A} = \frac{200\,ml}{Sample\,Volume\,(mL)}$$

Where, A is the volume of sample water and B is the volume of odor-free water added to make 200 mL of total water.

Specific conductivity of water: The total amount of dissolved salts present in water can be estimated by measuring the specific conductivity of water. The specific conductivity of water is determined by means of a portable ionic water tester and is expressed as micro-mho per cm at 25°C. 'mho' is the unit of conductivity and it equals to 1 Ampere per volt. The specific conductivity of water in micro mho per cm at 25°C is multiplied by a coefficient generally 0.65 so as to directly obtain the dissolved salt content in mg/L or ppm. The actual value of this coefficient depends upon the type of salt present in water.

Chemical Parameters

Solids: Total solids include suspended and dissolved solids. Amount of total solids in a water sample can be determined by evaporating the water and weighing the residue. Amount of suspended solids is determined by filtering the sample of water through filter paper, followed by drying the filter paper and weighing the solids. The quantity of dissolved solids including the colloidal solids is determined evaporating the filtered water (obtained from the suspended solid test) and weighing the residue.

Total solids can also be considered as the sum of organic and inorganic solids. Amount of inorganic solids can be determined by fusing the residue of total solids in a muffle-furnace and weighing the fused residue. Amount of organic solids is the difference between the amount of inorganic and total solid.

Hardness: Hardness of water is due to the presence of carbonates and sulphates of calcium and magnesium ions in the water. Sometimes hardness in the water can also be caused by the presence of chlorides and nitrates of calcium and magnesium.

Presence of hardness in water prevents the lathering of the soap during cleaning of clothes, etc.

Hardness is usually expressed in mg of calcium carbonate per litre of water. Hardness is generally determined by Versenate Method. In this method, the water is titrated against EDTA salt solution using Eriochrome Black T as indicator solution. While titrating, color changes from wine red to blue. In general, under a normal range of pH values, water with hardness up to 75 mg/L are considered as soft and those with 200 mg/L and above are considered as hard. In between, the water is considered as moderately hard. Underground water is generally harder than the surface water, as they have more opportunity to come in contact with minerals.

For boiler feed water and for efficient cloth washing, etc., the water must be soft. However, for drinking purposes, water with hardness below 75 mg/L is generally tasteless and hence, the prescribed hardness limit for drinking ranges between 75 to 150 mg/L.

Chlorides: Sodium chloride is the main substance in chloride water. The natural water near the mines and sea has dissolved sodium chloride. Similarly, the presence of chlorides may be due to the mixing of saline water and sewage in the water. Excess of chlorides is considered as dangerous and makes the water unfit for many uses.

Chloride content is determined by titrating the wastewater with silver nitrate and potassium chromate. Appearance of reddish color confirms presence of chlorides in water.

Chlorine: Dissolved free chlorine is never found in natural waters. It is present in the treated water resulting from disinfection with chlorine. The chlorine remains as residual in treated water for the sake of safety against pathogenic bacteria.

Residual chlorine is determined by the starch-iodide test. In starch-iodide test, potassium iodide and starch solutions are added to the sample of water due to which blue color is formed. This blue color is then removed by titrating with sodium thiosuplhate solution, and the quantity of chloride is calculated. On the addition of ortho-iodine solution if yellow color is formed, it indicates the presence of residual chlorine in the water. The intensity of this yellow color is compared with standard colors to determine the quantity of residual chlorine.

The residual chlorine should remain between 0.5 to 0.2 mg/L in the water so that it remains safe against pathogenic bacteria.

Iron and Manganese: These are generally found in ground water. The presence of iron and manganese in water makes it brownish red in color. Presence of these elements leads to the growth of micro-organism and corrodes the water pipes. Iron and manganese also causes taste and odor in the water. The quantity of iron and manganese is determined by colorimetric methods.

pH: pH value is the logarithm of reciprocal of hydrogen ion activity in moles per liter. Depending upon the nature of dissolved salts and minerals, water may be acidic or alkaline. When acids or alkalis are dissolved in water, they dissociate into electrically charged hydrogen and hydroxyl radicals, respectively. Dissolved gases such as carbon dioxide, hydrogen sulphide and ammonia also affect the pH of water. pH of natural water is generally in the range of 6-8. Industrial wastes may be strongly acidic or basic and their effect on pH value of receiving water depends on the buffering capacity of receiving water. pH lower than 4 have sour taste and above 8.5 have bitter taste. At pH below 6.5, corrosion starts to occur in pipes.

Lead and Arsenic: These are not usually found in natural waters. But sometimes lead is mixed up in water from lead pipes or from tanks lined with lead paint when water moves through them. These are poisonous and dangerous to the health of public. The presence of lead and arsenic is detected by means of chemical tests.

Dissolved Gases: Oxygen and carbon dioxide gases are found in the natural waters of all types. In addition, water may contain some amount of hydrogen sulphide and ammonia depending upon the pH and anaerobic/aerobic condition of water.

Surface water absorbs oxygen from the atmosphere. Algae and other tiny plant life of water also give oxygen to the water. Dissolved oxygen is necessary for sustenance of aquatic life in water and to keep it fresh. The water absorbs carbon dioxide from the atmosphere. Calcium and magnesium salts get converted into bicarbonates in presence of carbon dioxide and cause hardness in the water. The presence of carbon dioxide can easily determined by mixing the lime solution in the water.

Nitrogen: Nitrogen may be present in the water in the form of nitrites, nitrates, free ammonia, and albuminoidal nitrogen. The presence of nitrogen in the water indicates the presence of organic maters in the water.

The presence of the nitrites in the water, due to partly oxidized organic matters, is very dangerous. Therefore, in no case nitrites should be allowed in the water.

The nitrites are rapidly and easily converted to nitrates by the full oxidation of the organic matters. The presence of nitrates is not so harmful. But nitrates > 45 mg/L can cause "mathemoglobinemia" disease to the children.

Free ammonia is obtained from the decomposition of organic matters in the beginning, therefore if free ammonia is present in the water, it will indicate that the decomposition of the organic matters has started recently. The presence of nitrites indicates partial decomposition of organic matters, whereas the presence of nitrates indicates fully oxidized matters.

Metals and other chemical substances: Water contains various types of minerals and metals such as iron, manganese, copper, lead barium, cadmium, selenium, fluoride, arsenic, etc.

Arsenic, selenium are poisonous, therefore they must be removed totally. Human lungs are affected by the presence of high quantity of copper in the water. Fewer cavities in the teeth will be formed due to excessive presence of fluoride in water.

The quantity of the metals and other substances can be done indirectly by colorimetric methods using UV-visible spectrophotometer or directly by the use of sophisticated instruments such as Atomic Absorption Spectrophotometer (AAS), Atomic Emission Spectrophotometer (AES), Inductively Coupled Mass Spectrophotometer (ICP-MS), etc.

Alkalinity (A_T)

> ➤ Alkalinity is a measure of the ability of a solution to neutralize acids to the equivalence point of carbonate or bicarbonate. It is the water's ability to absorb hydrogen ions without significant pH change. Alkalinity is a measure of the buffering capacity of water.

> ➤ Alkalinity is equal to the stoichiometric sum of the bases in solution.

> ➤ In natural environment, carbonate alkalinity makes up most of the total alkalinity due to the common occurrence and dissolution of carbonate rocks and presence of carbon dioxide in the atmosphere.

> ➤ Other natural components that contribute to alkalinity include hydroxide, borate, phosphate, silicate, nitrate, dissolved ammonia, conjugate bases of some organic acids and sulfide.

> ➤ Alkalinity is usually expressed in meq/L (milliequivalent per liter).

$$Alkalinity\,(mol/L) = \left[HCO_3^-\right] + 2\left[CO_3^-\right] + \left[OH^-\right] - \left[H^+\right]$$

Where the quantities in parenthesis are concentrations in meq/L or mg/L as $CaCO_3$.

$$(mg/L)\,of\,X\,as\,CaCO_3 = \frac{Concentration\,of\,X\,(mg/L) \times 50\,mg\,CaCO_3\,/\,meq}{Equivalent\,weight\,of\,X\,(mg/meq)}$$

Biochemical Oxygen Demand (BOD)

> ➤ Biochemical Oxygen Demand (BOD) is a chemical procedure for determining how fast biological organisms use up oxygen in a body of water.

> ➤ It is used in water quality management and assessment, ecology and environmental science.

> ➤ BOD is not an accurate quantitative test, although it is considered as an indication of the quality of a water source.

> ➤ It is most commonly expressed in milligrams of oxygen consumed per litre of sample during 5 days of incubation at 20 °C or 3 days of incubation at 27 °C.

> ➤ The BOD test must be inhibited to prevent oxidation of ammonia. If the inhibitor is not added, the BOD will be between 10% and 40% higher than can be accounted for by carbonaceous oxidation.

Stages of Decomposition in the BOD test

> ➤ There are two stages of decomposition in the BOD test: a carbonaceous stage and a nitrogenous stage.

> ➤ The carbonaceous stage represents oxygen demand involved in the conversion of organic carbon to carbon dioxide.

> ➤ The second stage or the nitrogenous stage represents a combined carbonaceous plus nitrogenous demand, when organic nitrogen, ammonia and nitrite are converted to nitrate. Nitrogenous oxygen demand generally begins after about 6 days.

> ➤ Under some conditions, if ammonia, nitrite, and nitrifying bacteria are present, nitrification can occur in less than 5 days. In this case, a chemical compound that prevents nitrification is added to the sample if the intent is to measure only the carbonaceous demand. The results are reported as carbonaceous BOD (CBOD) or as $CBOD_5$ when a nitrification inhibitor is used.

BOD – Dilution Method: BOD is the amount of oxygen (Dissolved Oxygen (DO)) required for the biological decomposition of organic matter. The oxygen consumed is related to the amount of biodegradable organics.

When organic substances are broken down in water, oxygen is consumed

$$Organic\,Carbon + O_2 \rightarrow CO_2$$

Where, organic carbon in human waste includes protein, carbohydrates, fats, etc.

Measure of BOD = Initial oxygen- Final Oxygen after (5 days at 20°C) or (3 days at 27°C)

Two standard 300 mL BOD bottles are filled completely with wastewater. The bottles are sealed. Oxygen content (DO) of one bottle is determined immediately. The other bottle is incubated at 20°C for 5 days or (or at 27°C for 3 days) in total darkness to prevent algal growth. After which its oxygen content is again measured. The difference between the two DO values is the amount of oxygen consumed by micro-organisms during 5 days and is reported as BOD_5.

Since the saturated value of DO for water at 20°C is 9.1 mg/L only and that the oxygen demand for wastewater may be of the order of several hundred mg/L, therefore, wastewater are generally diluted so that the final DO in BOD test is always ≥ 2 mg/L. Precaution is also taken so as to obtain at least 2 mg/L change in DO between initial and final values.

$$BOD_5 = \frac{\left(DO_i - DO_f \right)}{P}$$

Where, DO_i and DO_f are initial and final DO concentrations of the diluted sample, respectively. P is called as dilution factor and it is the ratio of sample volume (volume of wastewater) to total volume (wastewater plus dilution water). In the above formula, it was assumed that the diluted wastewater had no oxygen demand of itself and that the dilution wastewater used was pure.

Most of the times, microorganisms are added in the dilution water so as to have enough microorganisms for carrying out biodegradation of organic waste. In this case, the oxygen demand of seeded water is subtracted from the demand of mixed sample of waste and dilution water. In this case,

$$BOD_5 = \frac{\left[(DO_i - DO_f) - (B_i - B_f)(1 - P)\right]}{P}$$

Where, B_i and B_f are initial and final DO concentrations of the seeded diluted water (blank).

Modeling BOD as First Order Reaction

Assuming that the rate of decomposition of organic waste is proportional to the waste left in the flask:

$$\frac{dL_t}{dt} = -kL_t$$

Where, L_t is the amount of oxygen demand left after time t and k is the BOD rate constant (time-1). Solving this equation yields

$$L_t = L_o e^{-kT}$$

Where, L_o is the ultimate carbonaceous oxygen demand and it is also the amount of O_2 demand left initially (at time 0, no DO demand has been exerted, so BOD = 0)

At any time, $L_o = BOD_t + L_t$ (that is the amount of DO demand used up and the amount of DO that could be used up eventually). Assuming that DO depletion is first order:

$$BOD_t = L_o \left(1 - e^{-kT}\right)$$

As temperature increases, metabolism increases, utilization of DO also increases, therefore, k is a function of temperature (T in °C). k at any temperature T (°C) is obtained as:

$$k_T = k_{20} \left(\theta\right)^{T-20}$$

Where, k_{20} is the value of k at 20°C and θ is an empirical constant. θ = 1.135 if T is between 4 - 20°C; θ = 1.056 if T is between 20 - 30°C.

Chemical Oxygen Demand (COD)

This test is carried out on the sewage to determine the extent of readily oxidizable organic matter, which is of two types:

a. Organic matter which can be biologically oxidized is called biologically active

b. Organic matter which cannot be oxidized biologically is called biologically inactive.

COD gives the oxygen required for the complete oxidation of both biodegradable and non-biodegradable matter.

➢ COD is a measure of the oxygen equivalent of the organic matter content of a sample that is susceptible to oxidation by a strong chemical oxidant.

➢ It is an indirect method to measure the amount of organic compounds in water.

➢ It is expressed in milligrams per liter (mg/L), which indicates the mass of oxygen consumed per liter of solution.

Analytical Procedure

$$Organic\ C + Cr_2O_7^- \rightarrow CO_2 + H_2O + Cr_2O_4^{2-}$$

➢ A sample is refluxed in strongly acidic solution with a known excess of potassium dichromate ($K_2Cr_2O_7$) for 2-3 h.

➢ After digestion, the remaining unreduced $K_2Cr_2O_7$ is titrated with ferrous ammonium sulphate to determine the amount of $K_2Cr_2O_7$ consumed.

➢ Then, the oxidizable matter is calculated in terms of oxygen equivalent.

➢ This procedure is applicable to COD values between 40 and 400 mg/L.

Essential Differences between BOD and COD

➢ COD always oxidize things that the BOD cannot or will not measure; therefore, COD is always higher than the BOD. The common compounds which cause COD to be higher than BOD include sulfides, sulfites, thiosulfates and chlorides.

➢ The general relationship between BOD and COD for sewage and most human wastes is about 1 unit of BOD≈0.64–0.68 units of COD. The relationship is not consistent and it may vary considerably for industrial wastewaters.

Fecal Indicator Bacteria

Fecal indicator bacteria, which are directly associated with fecal contamination, are used to detect the possible presence of waterborne pathogens by assessing the microbiological quality of water.

➢ Fecal material from warm-blooded animals may contain a variety of intestinal microorganisms (viruses, bacteria, and protozoa) that are pathogenic to humans. For example, bacterial pathogens of the Salmonella, Shigella and Vibrio can result in gastroenteritis and bacillary dysentery, typhoid fever, cholera, etc.

➢ The presence of E. coli in water is direct evidence of fecal contamination from warm- blooded animals.

➢ A few strains of E. coli are pathogenic, such as E. coli O157:H7, but most strains are not.

➢ Densities of other indicator bacteria (total coliforms, fecal coliforms, and fecal streptococci) can be, but are not necessarily, associated with fecal contamination.

➢ Despite this limitation, total coliforms are used to indicate ground-water susceptibility to fecal contamination. Fecal coliforms also are used as a measure of water safety for body- contact recreation or for consumption.

Usually, five types of fecal indicator bacteria i.e. total coliform bacteria, fecal coliform bacteria, Escherichia coli (E. coli), fecal streptococci, and enterococci are identified and quantified.

Following methods can be used to test for indicator bacteria:

➢ Total count of bacteria: In this method, total number of bacteria present in a milliliter of water is counted. The sample of water is diluted; 1 mL of sample water is diluted in 99 mL of sterilized water. Then 1 mL of diluted water is mixed with 10 mL of agar or gelatin (culture medium). This mixture is then kept in incubator at 37°C for 24 h or at 20°C for 48 h. After that, the sample is taken out from incubator and colonies of bacteria are counted by means of microscope. The product of the number of colonies and the dilution factor gives the total number of bacteria per mL of undiluted water sample.

➢ Membrane-filtration method: In this method, the sample is filtered through a sterilized membrane of special design due to which all bacteria get stained on the membrane. The member is then put in contact of culture medium in the incubator for 24 hours at 37°C. The membrane after incubation is taken out and the colonies of bacteria are counted by means of microscope.

➢ Liquid broth method, using the presence-absence format or the most-probable- number (MPN) format: In this method, the detection is done by mixing dilutions of a sample of water with lactose broth and keeping it in the incubator at for 48 h. The presence of acid or carbon dioxide gas in the test tube indicates presence of E-coli. After this, the standard statistical tables (Maccardy's) are referred and the 'Most Probable Number' (MPN) of E-coli per 100 mL of water is determined. MPN is the number which represents the bacterial density which is most likely to be present.

Complete Assessment of The Quality of The Aquatic Environment

➢ Chemical analyses of water and aquatic organisms

➢ Biological tests such as toxicity tests and measurements of enzyme activities

➢ Descriptions of aquatic organisms including their occurrence, density, biomass, physiology and diversity

➢ Physical measurements of water temperature, pH, conductivity, light penetration, particle size of suspended and dissolved material, flow velocity, hydrological balance, etc..

Following water quality parameters need to be determined to assess quality of water :

Dissolved oxygen	Usually decreases as discharge increases. Used as a water quality indicator in most water quality models.
Biochemical oxygen demand (BOD)	A measure of oxygen-reducing potential for waterborne discharges. Used in most water quality models.
Temperature	Often increased by discharges, especially from electric power plants. Relatively easy to model.
Ammonia nitrogen	Reduces dissolved oxygen concentrations and adds nitrate to water. Can be predicted by most water quality models.
Algal concentration	Increases with pollution, especially nitrates and phosphates. Predicted by moderately complex models.
Coliform bacteria	An indicator of contamination from sewage and animal waste
Nitrates	A nutrient for algal growth and a health hazard at very high concentrations in drinking water. Predicted by moderately complex models.
Phosphates	Nutrient for algal growth. Predicted by moderately complex models.
Toxic organic compounds	A wide variety of organic (carbon-based) compounds can affect aquatic life and may be directly hazardous to humans. Usually very difficult to model.
Heavy metals	Substances containing lead, mercury, cadmium, and other metals can cause both ecological and human health problems. Difficult to model in detail.

Table: Monitoring systems used to determine the quality of water
in water bodies and liquid effluents

Parameter	Sampling or monitoring system
General	
pH	pH meter ISO (1980–91), Water Quality Standards APHA, ASTM, BS, DIN, SCA
BOD	Determine dissolved oxygen concentration in the test solution before and after incubation (APHA, ASTM, BS, DIN, ISO, SCA); 40 CFR, Part 136; USEPA Method 405.1

COD	Digest with potassium dichromate in strong acid solution with silver sulfate as catalyst after sample homogenization (APHA, ASTM, BS, DIN, ISO, SCA); 40 CFR, Part 136; USEPA Method 410.1
AOX	USEPA Method 1650 (titrimetric)
TSS	Filtration 40 CFR, Part 136; USEPA Method 160.2; APHA, BS, DIN, ISO, SCA
Total dissolved solids (TDS)	Pretreatment with membrane filtration, followed by evaporation APHA, BS, DIN, ISO, SCA
Phenol	Extract with MIBK, followed by GC analysis USEPA Methods 420.1, 420.2
Sulfide	React with dimethlphenylenediamine and ferric chloride in acid solution to form methylene blue; USEPA Methods 376.1, 376.2
Oil and grease	Extract with light petroleum, evaporate solvent, and measure weight USEPA Method 413.1
Organic compounds	
Total organic carbon	UV oxidation followed by infrared analysis USEPA Method 415.1; APHA, ASTM, DIN, ISO, SCA
Organics	40 CFR, Part 136.3 (GC, GC/MS, HPLC, ASTM D4657-87)
PAHs	Gas chromatography with flame ionization detection
Pesticides	Gas chromatography; 40 CFR, Part 136.3, Table 1-D.
Inorganic substances	
General reference	40 CFR, Part 136.3, Table 1-B.
Metals	
Arsenic	Atomic absorption spectroscopy; APHA, ASTM, SCA
Cadmium	Atomic absorption spectrometry; APHA, ASTM, BS, DIN, ISO, SCA Inductively coupled plasma emission spectrometry; ASTM, DIN, SCA
Chromium	Atomic absorption spectrometry; APHA, ASTM, BS, DIN, ISO, SCA Inductively coupled plasma emission spectrometry; ASTM, DIN, SCA
Lead	Atomic absorption spectrometry; APHA, ASTM, BS, DIN, ISO, SCA Inductively coupled plasma emission spectrometry; ASTM, DIN, SCA
Mercury	Flameless atomic absorption spectrometry; APHA, ASTM, BS, DIN, ISO, SCA
Nickel	Atomic absorption spectrometry; APHA, ASTM, DIN, SCA Inductively coupled plasma emission spectrometry; ASTM, DIN, SCA
Zinc	Atomic absorption spectrometry; APHA, ASTM, BSI, DIN, ISO, SCA

Water Pollution

Water pollution is the contamination of water bodies (e.g. lakes, rivers, oceans, aquifers and groundwater). This form of environmental degradation occurs when pollutants are directly or indirectly discharged into water bodies without adequate treatment to remove harmful compounds.

Raw sewage and industrial waste in the New River as it passes from Mexicali to Calexico, California

Water pollution affects the entire biosphere – plants and organisms living in these bodies of water. In almost all cases the effect is damaging not only to individual species and population, but also to the natural biological communities.

Introduction

Pollution in the Lachine Canal, Canada

Water pollution is a major global problem which requires ongoing evaluation and revision of water resource policy at all levels (international down to individual aquifers and wells). It has been suggested that water pollution is the leading worldwide cause of deaths and diseases, and that it accounts for the deaths of more than 14,000 people daily. An estimated 580 people in India die of water pollution related illness every day. About 90 percent of the water in the cities of China is polluted. As of 2007, half a billion Chinese had no access to safe drinking water. In addition to the acute problems of water pollution in developing countries, developed countries also continue to struggle with pollution problems. For example, in the most recent national report on water quality in the United States, 44 percent of assessed stream miles, 64 percent of assessed lake acres, and 30 percent of assessed bays and estuarine square miles were classified as polluted. The head of China's national development agency said in 2007 that one quarter the length of China's seven main rivers were so poisoned the water harmed the skin.

Water is typically referred to as polluted when it is impaired by anthropogenic contaminants and either does not support a human use, such as drinking water, or undergoes a marked shift in its ability to support its constituent biotic communities, such as fish. Natural phenomena such as volcanoes, algae blooms, storms, and earthquakes also cause major changes in water quality and the ecological status of water.

Categories

Although interrelated, surface water and groundwater have often been studied and managed as separate resources. Surface water seeps through the soil and becomes groundwater. Conversely, groundwater can also feed surface water sources. Sources of surface water pollution are generally grouped into two categories based on their origin.

Point Sources

Point source pollution at a shipyard in Rio de Janeiro, Brazil

Point source water pollution refers to contaminants that enter a waterway from a single, identifiable source, such as a pipe or ditch. Examples of sources in this category include discharges from a sewage treatment plant, a factory, or a city storm drain. The U.S. Clean Water Act (CWA) defines point source for regulatory enforcement purposes. The CWA definition of point source was amended in 1987 to include municipal storm sewer systems, as well as industrial storm water, such as from construction sites.

Non-point Sources

Nonpoint source pollution refers to diffuse contamination that does not originate from a single discrete source. NPS pollution is often the cumulative effect of small amounts of contaminants gathered from a large area. A common example is the leaching out of nitrogen compounds from fertilized agricultural lands. Nutrient runoff in storm water from "sheet flow" over an agricultural field or a forest are also cited as examples of NPS pollution.

Blue drain and yellow fish symbol used by the UK Environment Agency to raise awareness of the ecological impacts of contaminating surface drainage

Contaminated storm water washed off of parking lots, roads and highways, called urban runoff, is sometimes included under the category of NPS pollution. However, because this runoff is typically channeled into storm drain systems and discharged through pipes to local surface waters, it becomes a point source.

Groundwater Pollution

Interactions between groundwater and surface water are complex. Consequently, groundwater pollution, also referred to as groundwater contamination, is not as easily classified as surface water pollution. By its very nature, groundwater aquifers are susceptible to contamination from sources that may not directly affect surface water bodies, and the distinction of point vs. non-point source may be irrelevant. A spill or ongoing release of chemical or radionuclide contaminants into soil (located away from a surface water body) may not create point or non-point source pollution but can contaminate the aquifer below, creating a toxic plume. The movement of the plume, called a plume front, may be analyzed through a hydrological transport model or groundwater model. Analysis of groundwater contamination may focus on soil characteristics and site geology, hydrogeology, hydrology, and the nature of the contaminants.

Causes

The specific contaminants leading to pollution in water include a wide spectrum of chemicals, pathogens, and physical changes such as elevated temperature and discoloration. While many of the chemicals and substances that are regulated may be naturally occurring (calcium, sodium, iron, manganese, etc.) the concentration is often the key in determining what is a natural component of water and what is a contaminant. High concentrations of naturally occurring substances can have negative impacts on aquatic flora and fauna.

Oxygen-depleting substances may be natural materials such as plant matter (e.g. leaves and grass) as well as man-made chemicals. Other natural and anthropogenic substances may cause turbidity (cloudiness) which blocks light and disrupts plant growth, and clogs the gills of some fish species.

Many of the chemical substances are toxic. Pathogens can produce waterborne diseases in either human or animal hosts. Alteration of water's physical chemistry includes acidity (change in pH), electrical conductivity, temperature, and eutrophication. Eutrophication is an increase in the concentration of chemical nutrients in an ecosystem to an extent that increases the primary productivity of the ecosystem. Depending on the degree of eutrophication, subsequent negative environmental effects such as anoxia (oxygen depletion) and severe reductions in water quality may occur, affecting fish and other animal populations.

Pathogens

HOW WATER IS CONTAMINATED

Poster to teach people in South Asia about human activities leading to the pollution of water sources

A manhole cover unable to contain a sanitary sewer overflow

Fecal sludge collected from pit latrines is dumped into a river at the Korogocho slum in Nairobi, Kenya

Disease-causing microorganisms are referred to as pathogens. Although the vast majority of bacteria are either harmless or beneficial, a few pathogenic bacteria can cause disease. Coliform bacteria, which are not an actual cause of disease, are commonly used as a bacterial indicator of water pollution. Other microorganisms sometimes found in surface waters that have caused human health problems include:

- *Burkholderia pseudomallei*

- *Cryptosporidium parvum*

- *Giardia lamblia*

- *Salmonella*

- *Norovirus* and other viruses

- *Parasitic worms including the Schistosoma type*

Muddy river polluted by sediment

High levels of pathogens may result from on-site sanitation systems (septic tanks, pit latrines) or inadequately treated sewage discharges. This can be caused by a sewage plant designed with less than secondary treatment (more typical in less-developed countries). In developed countries, older cities with aging infrastructure may have leaky sewage collection systems (pipes, pumps, valves), which can cause sanitary sewer overflows. Some cities also have combined sewers, which may discharge untreated sewage during rain storms.

Pathogen discharges may also be caused by poorly managed livestock operations.

Organic, Inorganic and Macroscopic Contaminants

Contaminants may include organic and inorganic substances.

A garbage collection boom in an urban-area stream in Auckland, New Zealand

Organic water pollutants include:

- Detergents

- Disinfection by-products found in chemically disinfected drinking water, such as chloroform

- Food processing waste, which can include oxygen-demanding substances, fats and grease

- Insecticides and herbicides, a huge range of organohalides and other chemical compounds

- Petroleum hydrocarbons, including fuels (gasoline, diesel fuel, jet fuels, and fuel oil) and lubricants (motor oil), and fuel combustion byproducts, from storm water runoff

- Volatile organic compounds, such as industrial solvents, from improper storage.

- Chlorinated solvents, which are dense non-aqueous phase liquids, may fall to the bottom of reservoirs, since they don't mix well with water and are denser.

    ○ Polychlorinated biphenyl (PCBs)

    ○ Trichloroethylene

- Perchlorate

- Various chemical compounds found in personal hygiene and cosmetic products

- Drug pollution involving pharmaceutical drugs and their metabolites

Inorganic water pollutants include:

- Acidity caused by industrial discharges (especially sulfur dioxide from power plants)

- Ammonia from food processing waste

- Chemical waste as industrial by-products

- Fertilizers containing nutrients--nitrates and phosphates—which are found in storm water runoff from agriculture, as well as commercial and residential use

- Heavy metals from motor vehicles (via urban storm water runoff) and acid mine drainage

- Secretion of creosote preservative into the aquatic ecosystem

- Silt (sediment) in runoff from construction sites, logging, slash and burn practices or land clearing sites.

Macroscopic pollution – large visible items polluting the water – may be termed "floatables" in an urban storm water context, or marine debris when found on the open seas, and can include such items as:

- Trash or garbage (e.g. paper, plastic, or food waste) discarded by people on the ground, along with accidental or intentional dumping of rubbish, that are washed by rainfall into storm drains and eventually discharged into surface waters

- Nurdles, small ubiquitous waterborne plastic pellets

- Shipwrecks, large derelict ships.

The Brayton Point Power Station in Massachusetts discharges heated water to Mount Hope Bay

Thermal Pollution

Thermal pollution is the rise or fall in the temperature of a natural body of water caused by human influence. Thermal pollution, unlike chemical pollution, results in a change in the physical properties of water. A common cause of thermal pollution is the use of water as a coolant by power plants and industrial manufacturers. Elevated water temperatures decrease oxygen levels, which can kill fish and alter food chain composition, reduce species biodiversity, and foster invasion by new thermophilic species. Urban runoff may also elevate temperature in surface waters.

Thermal pollution can also be caused by the release of very cold water from the base of reservoirs into warmer rivers.

Transport and Chemical reactions of Water Pollutants

Most water pollutants are eventually carried by rivers into the oceans. In some areas of the world the influence can be traced one hundred miles from the mouth by studies using hydrology transport models. Advanced computer models such as SWMM or the DSSAM Model have been used in many locations worldwide to examine the fate of pollutants in aquatic systems. Indicator filter-feeding species such as copepods have also been used to study pollutant fates in the New York Bight, for example. The highest toxin loads are not directly at the mouth of the Hudson River, but 100 km (62 mi) south, since several days are required for incorporation into planktonic tissue. The Hudson discharge flows south along the coast due to the coriolis force. Further south are areas of oxygen depletion caused by chemicals using up oxygen and by algae blooms, caused by excess nutrients from algal cell death and decomposition. Fish and shellfish kills have been reported, because toxins climb the food chain after small fish consume copepods, then large fish eat smaller fish, etc. Each successive step up the food chain causes a cumulative concentration of pollutants such as heavy metals (e.g. mercury) and persistent organic pollutants such as DDT. This is known as bio-magnification, which is occasionally used interchangeably with bio-accumulation.

A polluted river draining an abandoned copper mine on Anglesey

Large gyres (vortexes) in the oceans trap floating plastic debris. The North Pacific Gyre, for example, has collected the so-called "Great Pacific Garbage Patch", which is now estimated to be one hundred times the size of Texas. Plastic debris can absorb toxic chemicals from ocean pollution, potentially poisoning any creature that eats it. Many of these long-lasting pieces wind up in the stomachs of marine birds and animals. This results in obstruction of digestive pathways, which leads to reduced appetite or even starvation.

Many chemicals undergo reactive decay or chemical change, especially over long periods of time in groundwater reservoirs. A noteworthy class of such chemicals is the chlorinated hydrocarbons such as trichloroethylene (used in industrial metal degreasing and electronics manufacturing) and tetrachloroethylene used in the dry cleaning industry. Both of these chemicals, which are carcinogens themselves, undergo partial decomposition reactions, leading to new hazardous chemicals (including dichloroethylene and vinyl chloride).

Groundwater pollution is much more difficult to abate than surface pollution because groundwater can move great distances through unseen aquifers. Non-porous aquifers such as clays partially purify water of bacteria by simple filtration (adsorption and absorption), dilution, and, in some cases, chemical reactions and biological activity; however, in some cases, the pollutants merely transform to soil contaminants. Groundwater that moves through open fractures and caverns is not filtered and can be transported as easily as surface water. In fact, this can be aggravated by the human tendency to use natural sinkholes as dumps in areas of karst topography.

There are a variety of secondary effects stemming not from the original pollutant, but a derivative condition. An example is silt-bearing surface runoff, which can inhibit the penetration of sunlight through the water column, hampering photosynthesis in aquatic plants.

Measurement

Environmental scientists preparing water autosamplers

Water pollution may be analyzed through several broad categories of methods: physical, chemical and biological. Most involve collection of samples, followed by specialized analytical tests. Some methods may be conducted *in situ*, without sampling, such as temperature. Government agencies and research organizations have published standardized, validated analytical test methods to facilitate the comparability of results from disparate testing events.

Sampling

Sampling of water for physical or chemical testing can be done by several methods, depending on the accuracy needed and the characteristics of the contaminant. Many contamination events are sharply restricted in time, most commonly in association with rain events. For this reason "grab" samples are often inadequate for fully quantifying contaminant levels. Scientists gathering this type of data often employ auto-sampler devices that pump increments of water at either time or discharge intervals.

Sampling for biological testing involves collection of plants and/or animals from the surface water body. Depending on the type of assessment, the organisms may be identified for biosurveys (population counts) and returned to the water body, or they may be dissected for bioassays to determine toxicity.

Physical Testing

Common physical tests of water include temperature, solids concentrations (e.g., total suspended solids (TSS)) and turbidity.

Chemical Testing

Water samples may be examined using the principles of analytical chemistry. Many published test methods are available for both organic and inorganic compounds. Frequently used methods include pH, biochemical oxygen demand (BOD), chemical oxygen demand (COD), nutrients (nitrate and phosphorus compounds), metals (including copper, zinc, cadmium, lead and mercury), oil and grease, total petroleum hydrocarbons (TPH), and pesticides.

Biological Testing

Biological testing involves the use of plant, animal, and/or microbial indicators to monitor the health of an aquatic ecosystem. They are any biological species or group of species whose function, population, or status can reveal what degree of ecosystem or environmental integrity is present. One example of a group of bio-indicators are the copepods and other small water crustaceans that are present in many water bodies. Such organisms can be monitored for changes (biochemical, physiological, or behavioral) that may indicate a problem within their ecosystem.

Control of Pollution

Decisions on the type and degree of treatment and control of wastes, and the disposal and use of adequately treated wastewater, must be based on a consideration all the technical factors of each drainage basin, in order to prevent any further contamination or harm to the environment.

Sewage Treatment

Deer Island Wastewater Treatment Plant serving Boston, Massachusetts and vicinity

In urban areas of developed countries, domestic sewage is typically treated by centralized sewage treatment plants. Well-designed and operated systems (i.e., secondary treatment or better) can remove 90 percent or more of the pollutant load in sewage. Some plants have additional systems to remove nutrients and pathogens.

Cities with sanitary sewer overflows or combined sewer overflows employ one or more engineering approaches to reduce discharges of untreated sewage, including:

- utilizing a green infrastructure approach to improve storm water management capacity throughout the system, and reduce the hydraulic overloading of the treatment plant

- repair and replacement of leaking and malfunctioning equipment

- increasing overall hydraulic capacity of the sewage collection system (often a very expensive option).

A household or business not served by a municipal treatment plant may have an individual septic tank, which pre-treats the wastewater on site and infiltrates it into the soil.

Industrial Wastewater Treatment

Dissolved air flotation system for treating industrial wastewater

Some industrial facilities generate ordinary domestic sewage that can be treated by municipal facilities. Industries that generate wastewater with high concentrations of conventional pollutants (e.g. oil and grease), toxic pollutants (e.g. heavy metals, volatile organic compounds) or other non-conventional pollutants such as ammonia, need specialized treatment systems. Some of these facilities can install a pre-treatment system to remove the toxic components, and then send the partially treated wastewater to the municipal system. Industries generating large volumes of wastewater typically operate their own complete on-site treatment systems. Some industries have been successful at redesigning their manufacturing processes to reduce or eliminate pollutants, through a process called pollution prevention.

Heated water generated by power plants or manufacturing plants may be controlled with:

- cooling ponds, man-made bodies of water designed for cooling by evaporation, convection, and radiation

- cooling towers, which transfer waste heat to the atmosphere through evaporation and/or heat transfer

- cogeneration, a process where waste heat is recycled for domestic and/or industrial heating purposes.

Riparian buffer lining a creek in Iowa

Agricultural Wastewater Treatment

Non Point Source Controls

Sediment (loose soil) washed off fields is the largest source of agricultural pollution in the United States. Farmers may utilize erosion controls to reduce runoff flows and retain soil on their fields. Common techniques include contour plowing, crop mulching, crop rotation, planting perennial crops and installing riparian buffers.

Feedlot in the United States

Nutrients (nitrogen and phosphorus) are typically applied to farmland as commercial fertilizer, animal manure, or spraying of municipal or industrial wastewater (effluent) or sludge. Nutrients may also enter runoff from crop residues, irrigation water, wildlife, and atmospheric deposition. Farmers can develop and implement nutrient management plans to reduce excess application of nutrients and reduce the potential for nutrient pollution.

To minimize pesticide impacts, farmers may use Integrated Pest Management (IPM) techniques (which can include biological pest control) to maintain control over pests, reduce reliance on chemical pesticides, and protect water quality.

Point Source Wastewater Treatment

Farms with large livestock and poultry operations, such as factory farms, are called *concentrated animal feeding operations* or *feedlots* in the US and are being subject to increasing government regulation. Animal slurries are usually treated by containment in anaerobic lagoons before disposal by spray or trickle application to grassland. Constructed wetlands are sometimes used to facilitate treatment of animal wastes. Some animal slurries are treated by mixing with straw and composted at high temperature to produce a bacteriologically sterile and friable manure for soil improvement.

Erosion and Sediment Control from Construction Sites

Silt fence installed on a construction site

Sediment from construction sites is managed by installation of:

- erosion controls, such as mulching and hydroseeding, and

- sediment controls, such as sediment basins and silt fences.

Discharge of toxic chemicals such as motor fuels and concrete washout is prevented by use of:

- spill prevention and control plans, and

- specially designed containers (e.g. for concrete washout) and structures such as overflow controls and diversion berms.

Control of Urban Runoff (Storm Water)

Effective control of urban runoff involves reducing the velocity and flow of storm water, as well as reducing pollutant discharges. Local governments use a variety of storm water management techniques to reduce the effects of urban runoff. These techniques, called best management practices (BMPs) in the U.S., may focus on water quantity control, while others focus on improving water quality, and some perform both functions.

Retention basin for controlling urban runoff

Pollution prevention practices include low-impact development techniques, installation of green roofs and improved chemical handling (e.g. management of motor fuels & oil, fertilizers and pesticides). Runoff mitigation systems include infiltration basins, bioretention systems, constructed wetlands, retention basins and similar devices.

Thermal pollution from runoff can be controlled by storm water management facilities that absorb the runoff or direct it into groundwater, such as bioretention systems and infiltration basins. Retention basins tend to be less effective at reducing temperature, as the water may be heated by the sun before being discharged to a receiving stream.

Wastewater Quality Indicators

Wastewater quality indicators are laboratory test methodologies to assess suitability of wastewater for disposal or re-use. Tests selected and desired test results vary with the intended use or discharge location. Tests measure physical, chemical, and biological characteristics of the waste water.

Tests will determine the quality of this wastewater.

Physical Characteristics

Temperature

Aquatic organisms cannot survive outside of specific temperature ranges. Irrigation runoff and water cooling of power stations may elevate temperatures above the acceptable range for some species. Elevated temperature can also cause an algae bloom which reduces oxygen levels. Temperature may be measured with a calibrated thermometer.

Solids

Solid material in wastewater may be dissolved, suspended, or settled. Total dissolved solids or TDS (sometimes called filterable residue) is measured as the mass of residue remaining when a measured volume of filtered water is evaporated. The mass of dried solids remaining on the filter is called total suspended solids (TSS) or nonfiltrable residue. Settleable solids are measured as the visible volume accumulated at the bottom of an Imhoff cone after water has settled for one hour. Turbidity is a measure of the light scattering ability of suspended matter in the water. Salinity measures water density or conductivity changes caused by dissolved materials.

Chemical Characteristics

Virtually any chemical may be found in water, but routine testing is commonly limited to a few chemical elements of unique significance.

Hydrogen

Water ionizes into hydronium (H_3O) cations and hydroxyl (OH) anions. The concentration of ionized hydrogen (as protonated water) is expressed as pH.

Oxygen

Most aquatic habitats are occupied by fish or other animals requiring certain minimum dissolved oxygen concentrations to survive. Dissolved oxygen concentrations may be measured directly in wastewater, but the amount of oxygen potentially required by other chemicals in the wastewater is termed an oxygen demand. Dissolved or suspended oxidizable organic material in wastewater will be used as a food source. Finely divided material is readily available to microorganisms whose populations will increase to digest the amount of food available. Digestion of this food requires oxygen, so the oxygen content of the water will ultimately be decreased by the amount required to digest the dissolved or suspended food. Oxygen concentrations may fall below the minimum required by aquatic animals if the rate of oxygen utilization exceeds replacement by atmospheric oxygen.

The reaction for biochemical oxidation may be written as:

Oxidizable material + bacteria + nutrient + $O_2 \rightarrow CO_2 + H_2O$ + oxidized inorganics such as NO_3^- or SO_4^{2-}

Oxygen consumption by reducing chemicals such as sulfides and nitrites is typified as follows:

$$S^{2-} + 2O_2 \rightarrow SO_4^{2-}$$
$$NO_2^- + \frac{1}{2}O_2 \rightarrow NO_3^-$$

Since all natural waterways contain bacteria and nutrient, almost any waste compounds introduced into such waterways will initiate biochemical reactions. Those biochemical reactions create what is measured in the laboratory as the biochemical oxygen demand (BOD).

Oxidizable chemicals (such as reducing chemicals) introduced into a natural water will similarly initiate chemical reactions (such as shown above). Those chemical reactions create what is measured in the laboratory as the chemical oxygen demand (COD).

Both the BOD and COD tests are a measure of the relative oxygen-depletion effect of a waste contaminant. Both have been widely adopted as a measure of pollution effect. The BOD test measures the oxygen demand of biodegradable pollutants whereas the COD test measures the oxygen demand of biodegradable pollutants plus the oxygen demand of non-biodegradable oxidizable pollutants.

The so-called 5-day BOD measures the amount of oxygen consumed by biochemical oxidation of waste contaminants in a 5-day period. The total amount of oxygen consumed when the biochemical reaction is allowed to proceed to completion is called the Ultimate BOD. The Ultimate BOD is too time consuming, so the 5-day BOD has almost universally been adopted as a measure of relative pollution effect.

There are also many different COD tests. Perhaps, the most common is the 4-hour COD.

There is no generalized correlation between the 5-day BOD and the Ultimate BOD. Likewise, there is no generalized correlation between BOD and COD. It is possible to develop such correlations for a specific waste contaminant in a specific wastewater stream, but such correlations cannot be generalized for use with any other waste contaminants or wastewater streams.

The laboratory test procedures for determining the above oxygen demands are detailed in the following sections of the "Standard Methods For the Examination of Water and Wastewater":

- 5-day BOD and Ultimate BOD: Sections 5210B and 5210C

- COD: Section 5220

Nitrogen

Nitrogen is an important nutrient for plant and animal growth. Atmospheric nitrogen is less biologically available than dissolved nitrogen in the form of ammonia and nitrates. Availability of dissolved nitrogen may contribute to algal blooms. Ammonia and organic forms of nitrogen are often measured as Total Kjeldahl Nitrogen, and analysis for inorganic forms of nitrogen may be performed for more accurate estimates of total nitrogen content.

Phosphates

Total Phosphorus and Phosphate, PO_4^{3-}

Phosphates enter the water ways through both non-point sources and point sources. Non-point source (NPS) pollution refers to water pollution from diffuse sources. Non-point source pollution can be contrasted with point source pollution, where discharges occur to a body of water at a single location. The non-point sources of phosphates include: natural decomposition of rocks and minerals, storm water runoff, agricultural runoff, erosion and sedimentation, atmospheric deposition, and direct input by animals/wildlife; whereas: point sources may include: waste water treatment plants and permitted industrial discharges. In general, the non-point source pollution typically is significantly higher than the point sources of pollution. Therefore, the key to sound management is to limit the input from both point and non-point sources of phosphate. High concentration of phosphate in water bodies is an indication of pollution and largely responsible for eutrophication.

Phosphates are not toxic to people or animals unless they are present in very high levels. Digestive problems could occur from extremely high levels of phosphate.

The following criteria for total phosphorus were recommended by the U.S. Environmental Protection Agency.

1. No more than 0.1 mg/L for streams which do not empty into reservoirs,

2. No more than 0.05 mg/L for streams discharging into reservoirs, and

3. No more than 0.025 mg/L for reservoirs.

Phosphorus is normally low (< 1 mg/l) in clean potable water sources and usually not regulated;

Chlorine

Chlorine has been widely used for bleaching, as a disinfectant, and for biofouling prevention in water cooling systems. Remaining concentrations of oxidizing hypochlorous acid and hypochlorite ions may be measured as chlorine residual to estimate effectiveness of disinfection or to demonstrate safety for discharge to aquatic ecosystems.

Biological Characteristics

Water may be tested by a bioassay comparing survival of an aquatic test species in the wastewater in comparison to water from some other source. Water may also be evaluated to determine the approximate biological population of the wastewater. Pathogenic micro-organisms using water as a means of moving from one host to another may be present in sewage. Coliform index measures the population of an organism commonly found in the intestines of warm-blooded animals as an indicator of the possible presence of other intestinal pathogens.

Water Treatment

Dalecarlia Water Treatment Plant, Washington, D.C.

Water treatment is any process that makes water more acceptable for a specific end-use. The end use may be drinking, industrial water supply, irrigation, river flow maintenance, water recreation or many other uses, including being safely returned to the environment. Water treatment removes contaminants and undesirable components, or reduces their concentration so that the water becomes fit for its desired end-use.

Treatment for Drinking Water Production

Treatment for drinking water production involves the removal of contaminants from raw water to produce water that is pure enough for human consumption without any short term or long term risk of any adverse health effect. Substances that are removed during the process of drinking water treatment include suspended solids, bacteria, algae, viruses, fungi, and minerals such as iron and manganese.

The processes involved in removing the contaminants include physical processes such as settling and filtration, chemical processes such as disinfection and coagulation and biological processes such as slow sand filtration.

Measures taken to ensure water quality not only relate to the treatment of the water, but to its conveyance and distribution after treatment. It is therefore common practice to keep residual disinfectants in the treated water to kill bacteriological contamination during distribution.

World Health Organization (WHO) guidelines are a general set of standards intended to apply where better local standards are not implemeted. More rigorous standards apply across Europe, the USA and in most other developed countries. followed throughout the world for drinking water quality requirements.

Processes

Empty aeration tank for iron precipitation

Tanks with sand filters to remove precipitated iron (not working at the time)

A combination selected from the following processes is used for municipal drinking water treatment worldwide:

- Pre-chlorination for algae control and arresting biological growth

- Aeration along with pre-chlorination for removal of dissolved iron when present with small amounts relatively of manganese

- Coagulation for flocculation or slow-sand filtration

- Coagulant aids, also known as polyelectrolytes – to improve coagulation and for more robust floc formation

- Sedimentation for solids separation that is removal of suspended solids trapped in the floc

- Filtration to remove particles from water either by passage through a sand bed that can be washed and reused or by passage through a purpose designed filter that may be washable.

- Disinfection for killing bacteria viruses and other pathogens.

Technologies for potable water and other uses are well developed, and generalized designs are available from which treatment processes can be selected for pilot testing on the specific source water. In addition, a number of private companies provide patented technological solutions for treatment of specific contaminants. Automation of water and waste-water treatment is common in the developed world. Source water quality through the seasons, scale and environmental impact can dictate capital costs and operating costs. End use of the treated water dictates the necessary quality monitoring technologies, and locally available skills typically dictate the level of automation adopted.

Constituent	Unit Processes
Turbidity and particles	Coagulation/ flocculation, sedimentation, granular filtration
Major dissolved inorganics	Softening, aeration, membranes
Minor dissolved inorganics	Membranes
Pathogens	Sedimentation, filtration, disinfection
Major dissolved organics	Membranes, adsorption

Polluted Water Treatment

A sewage treatment plant in northern Portugal

Wastewater treatment is the process that removes the majority of the contaminants from wastewater or sewage and produces both a liquid effluent suitable for disposal to the natural environment and a sludge. Biological processes can be employed in the treatment of wastewater and these processes may include, for example, aerated lagoons, activated sludge or slow sand filters. To be effective, sewage must be conveyed to a treatment plant by appropriate pipes and infrastructure and the process itself must be subject to regulation and controls. Some wastewaters require different and sometimes specialized treatment methods. At the simplest level, treatment of sewage and most wastewaters is carried out through separation of solids from liquids, usually by sedimentation. By progressively converting dissolved material into solids, usually a biological floc, which is then settled out, an effluent stream of increasing purity is produced.

Industrial Water and Wastewater Treatment

Two of the main processes of industrial water treatment are *boiler water treatment* and *cooling water treatment*. A lack of proper water treatment can lead to the reaction of solids and bacteria within pipe work and boiler housing. Steam boilers can suffer from scale or corrosion when left untreated. Scale deposits can lead to weak and dangerous machinery, while additional fuel is required to heat the same level of water because of the rise in thermal resistance. Poor quality dirty water can become a breeding ground for bacteria such as *Legionella* causing a risk to public health.

With the proper treatment, a significant proportion of industrial on-site wastewater might be reusable. This can save money in three ways: lower charges for lower water consumption, lower charges for the smaller volume of effluent water discharged and lower energy costs due to the recovery of heat in recycled wastewater.

Corrosion in low pressure boilers can be caused by dissolved oxygen, acidity and excessive alkalinity. Water treatment therefore should remove the dissolved oxygen and maintain the boiler water with the appropriate pH and alkalinity levels. Without effective water treatment, a cooling water system can suffer from scale formation, corrosion and fouling and may become a breeding ground for harmful bacteria. This reduces efficiency, shortens plant life and makes operations unreliable and unsafe.

Domestic Water Treatment

Water supplied to domestic properties may be further treated before use, often using an in-line treatment process. Such treatments can include water softening or ion exchange. Many propriety systems also claim to remove residual disinfectants and heavy metal ions.

Desalination

Saline water can be treated to yield fresh water. Two main processes are used, reverse osmosis or distillation. Both methods require high energy inputs and are usually only used where fresh water is difficult to source.

Field Processes

Living away from drinking water supplies often requires some form of portable water treatment process. These can vary in complexity from the simple addition of a disinfectant tablet in a hiker's water bottle through to complex multi-stage processes carried by boat or plane to disaster areas.

Ultra Pure Water Production

Some industries such as the production of silicon wafers, space technology and many high quality metallurgical process require ultrapure water. The production of such water typically involves many stages, and can include reverse osmosis, ion exchange and several distillation stages using solid tin apparatus.

History

Early water treatment methods still used included sand filtration and chlorination. The first documented use of sand filters to purify the water supply dates to 1804, when the owner of a bleachery in Paisley, Scotland, John Gibb, installed an experimental filter, selling his unwanted surplus to the public. This method was refined in the following two decades, and it culminated in the first treated public water supply in the world, installed by the Chelsea Waterworks Company in London in 1829.

Society and Culture

Developing Countries

As of 2006, waterborne diseases are estimated to have caused 1.8 million deaths each year. These deaths are attributable to inadequate public sanitation systems and in these cases, proper sewerage (or other options such as small-scale wastewater treatment) that must be installed.

Appropriate technology options in water treatment include both community-scale and household-scale point-of-use (POU) designs. Such designs may employ solar water disinfection methods, using solar irradiation to inactivate harmful waterborne microorganisms directly, mainly by the UV-A component of the solar spectrum, or indirectly through the presence of an oxide photocatalyst, typically supported TiO_2 in its anatase or rutile phases. Despite progress in SODIS technology, military surplus water treatment units like the ERDLator are still frequently used in developing countries. Newer military style Reverse Osmosis Water Purification Units (ROWPU) are portable, self-contained water treatment plants are becoming more available for public use.

For waterborne disease reduction to last, water treatment programs that research and development groups start in developing countries must be sustainable by the citizens of those countries. This can ensure the efficiency of such programs after the departure of the research team, as monitoring is difficult because of the remoteness of many locations.

Energy Consumption

For many cities, drinking water and wastewater treatment plants are typically the largest energy consumers, having a total of 30-40% of the cities' energy consumption. More than 4% of the nation's electricity goes towards moving and treating water and wastewater. Cost of these energy is consumed in the flocculation basin for drinking water treatment plants and in the aeration basin for wastewater treatment plants. High amount of energy is needed to mix the large volume of water to allow sedimentations to flocculate together. There are current technologies that may aim to reduce this amount of energy. These include optimizing system processes by modifying and improving pumping and aeration equipments.The effectiveness of such technologies are still under discussion as they take up a lot of energy.

Structure and Building of Water Treatment Plants

Economies of scale favor a large water treatment plant that small ones. There are various types of waste water treatment plants including sewage treatment plants, tertiary treatment plants, industrial waste water treatment plant, agriculture waste water treatment plant and leachate treatment plant. The largest feature of drinking water treatment plant is the clarifier and each tank is approximately 20 feet deep and 117 feet square. Most water treatment plants also have a row of filtration tanks which are 13 feet deep and 77 feet long, almost the length of an Olympic size poo l. These are the major components of any water treatment plants and take up the most space for all water treatment plants.This is a clear example of efficiency as this new technology allows much energy and manpower savings by treating wastewater directly to potable water. With new scientific advancements, we can look forward to newer and improved technology which makes our society run more efficiently. It may be thought to be more equitable to have smaller water treatment plants in every city. However, this is ineffective and a waste of resources as the economies of scale allows for great cost savings.

Notable Examples

A notable example that combines both wastewater treatment and drinking water treatment is NEWater in Singapore. NEWater is a technology practised in Singapore that converts wastewater to potable water. More specifically, it is treated wastewater (sewage) that has been purified using dual-membrane (via microfiltration and reverse osmosis) and ultraviolet technologies, in addition to conventional water treatment processes. The water is potable and is consumed by humans, but is mostly used by industries requiring high purity water. The total capacity of the plants is about 20 million US gallons per day (75,700 m^3/day). Some 6% of this is used for indirect potable use, equal to about 1% of Singapore's potable water requirement of 380 million US gallons per day (13 m^3/s). The rest is used at wafer fabrication plants and other non-potable applications in industries in Woodlands, Tampines, Pasir Ris, and Ang Mo Kio.

Failures of Water Treatment Plants

When water treatment plants fail, the impact reaches a large group of people. These water treatment plants may fail due to a variety of reasons. They include poor maintenance, power shutdown or the plant may simply not be able to withstand and treat such a high influx of water. This is why engineers often employ safe margin and design for a larger volume than expected. There are a few ways in which citizens can deal with a water treatment plant failure. This includes buying water filtration systems such as those from Brita or water filtration tablets.

Regulation by the US Government

Drinking Water

The Safe Drinking Water Act requires the U.S. Environmental Protection Agency (EPA) to set standards for drinking water quality in public water systems (entities that provide water for human consumption to at least 25 people for at least 60 days a year). Enforcement of the standards is mostly carried out by state health agencies. States may set standards that are more stringent than the federal standards.

EPA has set standards for over 90 contaminants organized into six groups: microorganisms, disinfectants, disinfection byproducts, inorganic chemicals, organic chemicals and radionuclides.

EPA also identifies and lists unregulated contaminants which may require regulation. The *Contaminant Candidate List* is published every five years, and EPA is required to decide whether to regulate at least five or more listed contaminants.

Local drinking water utilities may apply for low interest loans, to make facility improvements, through the Drinking Water State Revolving Fund.

Wastewater

EPA and state environmental agencies set wastewater standards under the Clean Water Act. Point sources must obtain surface water discharge permits through the National Pollutant Discharge Elimination System (NPDES).

EPA sets basic national wastewater standards:

- The "Secondary Treatment Regulation" for municipal sewage treatment plants, and

- Effluent guidelines for categories of industrial facilities.

These standards are incorporated into the permits, which may include additional treatment requirements developed on a case-by-case basis. NPDES permits must be renewed every five years. EPA has authorized 46 state agencies to issue and enforce NPDES permits. EPA regional offices issues permits for the rest of the country.

Financial assistance for improvements to sewage treatment facilities is available to state and local governments through the Clean Water State Revolving Fund, a low interest loan program.

Sewage Treatment

Wastewater treatment plant in Massachusetts, United States

Sewage treatment is the process of removing contaminants from wastewater, primarily from household sewage. It includes physical, chemical, and biological processes to remove these contaminants and produce environmentally safer treated wastewater (or treated effluent). A by-product of sewage treatment is usually a semi-solid waste or slurry, called sewage sludge, that has to undergo further treatment before being suitable for disposal or land application.

Sewage treatment may also be referred to as wastewater treatment, although the latter is a broader term which can also be applied to purely industrial wastewater. For most cities, the sewer system will also carry a proportion of industrial effluent to the sewage treatment plant which has usually received pretreatment at the factories themselves to reduce the pollutant load. If the sewer system is a combined sewer then it will also carry urban runoff (stormwater) to the sewage treatment plant. Sewage water can travel towards treatment plants via piping and in a flow aided by gravity and pumps. The first part of filtration of sewage typically includes a bar screen to filter solids and large objects which are then collected in dumpsters and disposed of in landfills. Fat and grease will also be removed before the primary treatment of sewage.

Terminology

The term "sewage treatment plant" (or "sewage treatment works" in some countries) is nowadays often replaced with the term "wastewater treatment plant".

Sewage can be treated close to where the sewage is created, which may be called a "decentralized" system or even an "on-site" system (in septic tanks, biofilters or aerobic treatment systems). Alternatively, sewage can be collected and transported by a network of pipes and pump stations to a municipal treatment plant. This is called a "centralized" system.

Origins of Sewage

Sewage is generated by residential, institutional, commercial and industrial establishments. It includes household waste liquid from toilets, baths, showers, kitchens, and sinks draining into sewers. In many areas, sewage also includes liquid waste from industry and commerce. The separation and draining of household waste into greywater and blackwater is becoming more common in the developed world, with treated greywater being permitted to be used for watering plants or recycled for flushing toilets.

Sewage Mixing with Rainwater

Sewage may include stormwater runoff or urban runoff. Sewerage systems capable of handling storm water are known as combined sewer systems. This design was common when urban sewerage systems were first developed, in the late 19th and early 20th centuries. Combined sewers require much larger and more expensive treatment facilities than sanitary sewers. Heavy volumes of storm runoff may overwhelm the sewage treatment system, causing a spill or overflow. Sanitary sewers are typically much smaller than combined sewers, and they are not designed to transport stormwater. Backups of raw sewage can occur if excessive infiltration/inflow (dilution by stormwater and/or groundwater) is allowed into a sanitary sewer system. Communities that have urbanized in the mid-20th century or later generally have built separate systems for sewage (sanitary sewers) and stormwater, because precipitation causes widely varying flows, reducing sewage treatment plant efficiency.

As rainfall travels over roofs and the ground, it may pick up various contaminants including soil particles and other sediment, heavy metals, organic compounds, animal waste, and oil and grease. Some jurisdictions require stormwater to receive some level of treatment before being discharged directly into waterways. Examples of treatment processes used for stormwater include retention basins, wetlands, buried vaults with various kinds of media filters, and vortex separators (to remove coarse solids).

Industrial Effluent

In highly regulated developed countries, industrial effluent usually receives at least pretreatment if not full treatment at the factories themselves to reduce the pollutant load, before discharge to the sewer. This process is called industrial wastewater treatment. The same does not apply to many developing countries where industrial effluent is more likely to enter the sewer if it exists, or even the receiving water body, without pretreatment.

Industrial wastewater may contain pollutants which cannot be removed by conventional sewage treatment. Also, variable flow of industrial waste associated with production cycles may upset the population dynamics of biological treatment units, such as the activated sludge process.

Process Steps

Sewage collection and treatment is typically subject to local, state and federal regulations and standards.

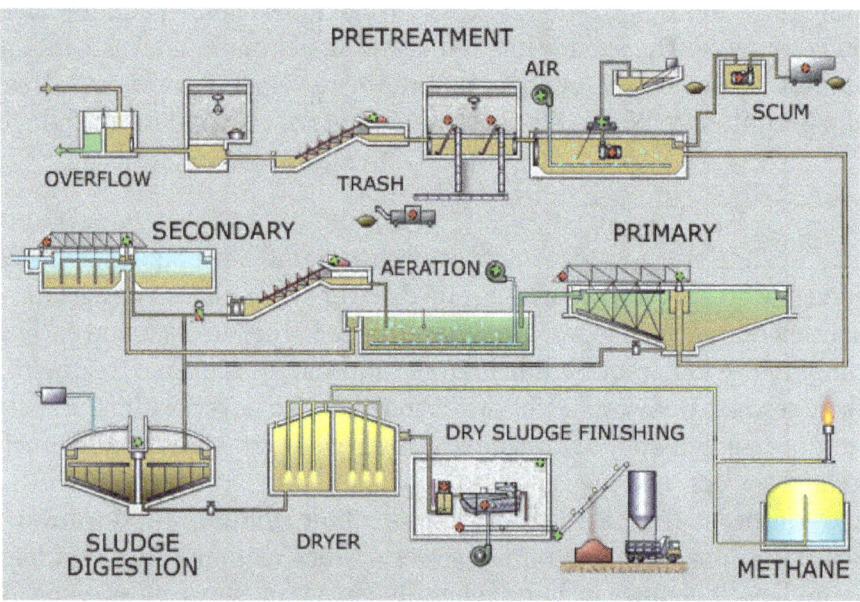

Simplified process flow diagram for a typical large-scale treatment plant

Treating wastewater has the aim to produce an effluent that will do as little harm as possible when discharged to the surrounding environment, thereby preventing pollution compared to releasing untreated wastewater into the environment.

Sewage treatment generally involves three stages, called primary, secondary and tertiary treatment.

- *Primary treatment* consists of temporarily holding the sewage in a quiescent basin where heavy solids can settle to the bottom while oil, grease and lighter solids float to the surface. The settled and floating materials are removed and the remaining liquid may be discharged or subjected to secondary treatment. Some sewage treatment plants that are connected to a combined sewer system have a bypass arrangement after the primary treatment unit. This means that during very heavy rainfall events, the secondary and tertiary treatment systems can be bypassed to protect them from hydraulic overloading, and the mixture of sewage and stormwater only receives primary treatment.

- *Secondary treatment* removes dissolved and suspended biological matter. Secondary treatment is typically performed by indigenous, water-borne micro-organisms in a managed habitat. Secondary treatment may require a separation process to remove the micro-organisms from the treated water prior to discharge or tertiary treatment.

- *Tertiary treatment* is sometimes defined as anything more than primary and secondary treatment in order to allow rejection into a highly sensitive or fragile ecosystem (estuaries, low-flow rivers, coral reefs,...). Treated water is sometimes disinfected chemically or physically (for example, by lagoons and microfiltration) prior to discharge into a stream, river, bay, lagoon or wetland, or it can be used for the irrigation of a golf course, green way or park. If it is sufficiently clean, it can also be used for groundwater recharge or agricultural purposes.

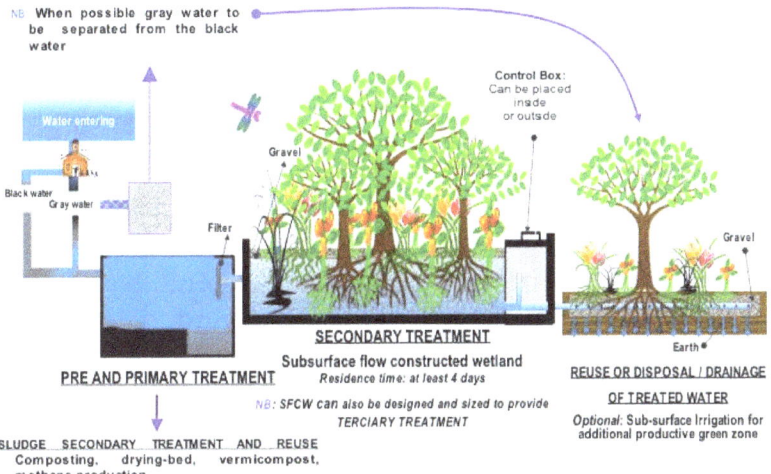

Process flow diagram for a typical treatment plant via subsurface flow constructed wetlands (SFCW)

Pretreatment

Pretreatment removes all materials that can be easily collected from the raw sewage before they damage or clog the pumps and sewage lines of primary treatment clarifiers. Objects commonly removed during pretreatment include trash, tree limbs, leaves, branches, and other large objects.

The influent in sewage water passes through a bar screen to remove all large objects like cans, rags, sticks, plastic packets etc. carried in the sewage stream. This is most commonly done with an automated mechanically raked bar screen in modern plants serving large populations, while in smaller or less modern plants, a manually cleaned screen may be used. The raking action of a mechanical bar screen is typically paced according to the accumulation on the bar screens and/or flow rate. The solids are collected and later disposed in a landfill, or incinerated. Bar screens or mesh screens of varying sizes may be used to optimize solids removal. If gross solids are not removed, they become entrained in pipes and moving parts of the treatment plant, and can cause substantial damage and inefficiency in the process.

Grit Removal

Pretreatment may include a sand or grit channel or chamber, where the velocity of the incoming sewage is adjusted to allow the settlement of sand, grit, stones, and broken glass. These particles are removed because they may damage pumps and other equipment. For small sanitary sewer systems, the grit chambers may not be necessary, but grit removal is desirable at larger plants. Grit chambers come in 3 types: horizontal grit chambers, aerated grit chambers and vortex grit chambers. The process is called sedimentation.

Flow Equalization

Clarifiers and mechanized secondary treatment are more efficient under uniform flow conditions. Equalization basins may be used for temporary storage of diurnal or wet-weather flow peaks. Basins provide a place to temporarily hold incoming sewage during plant maintenance and a means of diluting and distributing batch discharges of toxic or high-strength waste which might otherwise inhibit biological secondary treatment (including portable toilet waste, vehicle holding tanks, and septic tank pumpers). Flow equalization basins require variable discharge control, typically include provisions for bypass and cleaning, and may also include aerators. Cleaning may be easier if the basin is downstream of screening and grit removal.

Fat and Grease Removal

In some larger plants, fat and grease are removed by passing the sewage through a small tank where skimmers collect the fat floating on the surface. Air blowers in the base of the tank may also be used to help recover the fat as a froth. Many plants, however, use primary clarifiers with mechanical surface skimmers for fat and grease removal.

Primary Treatment

Primary treatment tanks in Oregon, USA.

In the primary sedimentation stage, sewage flows through large tanks, commonly called "pre-settling basins", "primary sedimentation tanks" or "primary clarifiers". The tanks are used to settle sludge while grease and oils rise to the surface and are skimmed off. Primary settling tanks are usually equipped with mechanically driven scrapers that continually drive the collected sludge towards a hopper in the base of the tank where it is pumped to sludge treatment facilities. Grease and oil from the floating material can sometimes be recovered for saponification (soap making).

Secondary Treatment

Secondary treatment is designed to substantially degrade the biological content of the sewage which are derived from human waste, food waste, soaps and detergent. The majority of municipal plants treat the settled sewage liquor using aerobic biological processes. To be effective, the biota require both oxygen and food to live. The bacteria and protozoa consume biodegradable soluble organic contaminants (e.g. sugars, fats, organic short-chain carbon molecules, etc.) and bind much of the less soluble fractions into floc. Secondary treatment systems are classified as *fixed-film* or *suspended-growth* systems.

- Fixed-film or attached growth systems include trickling filters, bio-towers, and rotating biological contactors, where the biomass grows on media and the sewage passes over its surface. The fixed-film principle has further developed into Moving Bed Biofilm Reactors (MBBR) and Integrated Fixed-Film Activated Sludge (IFAS) processes. An MBBR system typically requires a smaller footprint than suspended-growth systems.

- Suspended-growth systems include activated sludge, where the biomass is mixed with the sewage and can be operated in a smaller space than trickling filters that treat the same amount of water. However, fixed-film systems are more able to cope with drastic changes in the amount of biological material and can provide higher removal rates for organic material and suspended solids than suspended growth systems.

Secondary Sedimentation

Secondary clarifier at a rural treatment plant.

Some secondary treatment methods include a secondary clarifier to settle out and separate biological floc or filter material grown in the secondary treatment bioreactor.

List of Process Types

- Activated sludge
- Aerated lagoon
- Aerobic granulation
- Constructed wetland
- Membrane bioreactor
- Rotating biological contactor
- Sequencing batch reactor
- Trickling filter

To use less space, treat difficult waste, and intermittent flows, a number of designs of hybrid treatment plants have been produced. Such plants often combine at least two stages of the three main treatment stages into one combined stage. In the UK, where a large number of wastewater treatment plants serve small populations, package plants are a viable alternative to building a large structure for each process stage. In the US, package plants are typically used in rural areas, highway rest stops and trailer parks.

Tertiary Treatment

The purpose of tertiary treatment is to provide a final treatment stage to further improve the effluent quality before it is discharged to the receiving environment (sea, river, lake, wet lands, ground, etc.). More than one tertiary treatment process may be used at any treatment plant. If disinfection is practised, it is always the final process. It is also called "effluent polishing."

Filtration

Sand filtration removes much of the residual suspended matter. Filtration over activated carbon, also called *carbon adsorption,* removes residual toxins.

Lagoons or Ponds

Lagoons or ponds provide settlement and further biological improvement through storage in large man-made ponds or lagoons. These lagoons are highly aerobic and colonization by native macrophytes, especially reeds, is often encouraged. Small filter-feeding invertebrates such as *Daphnia* and species of *Rotifera* greatly assist in treatment by removing fine particulates.

A sewage treatment plant and lagoon in Everett, Washington, United States.

Biological Nutrient Removal

Biological nutrient removal (BNR) is regarded by some as a type of secondary treatment process, and by others as a tertiary (or "advanced") treatment process.

Wastewater may contain high levels of the nutrients nitrogen and phosphorus. Excessive release to the environment can lead to a buildup of nutrients, called eutrophication, which can in turn encourage the overgrowth of weeds, algae, and cyanobacteria (blue-green algae). This may cause an algal bloom, a rapid growth in the population of algae. The algae numbers are unsustainable and eventually most of them die. The decomposition of the algae by bacteria uses up so much of the oxygen in the water that most or all of the animals die, which creates more organic matter for the bacteria to decompose. In addition to causing deoxygenation, some algal species produce toxins that contaminate drinking water supplies. Different treatment processes are required to remove nitrogen and phosphorus.

Nitrogen Removal

Nitrogen is removed through the biological oxidation of nitrogen from ammonia to nitrate (nitrification), followed by denitrification, the reduction of nitrate to nitrogen gas. Nitrogen gas is released to the atmosphere and thus removed from the water.

Nitrification itself is a two-step aerobic process, each step facilitated by a different type of bacteria. The oxidation of ammonia (NH_3) to nitrite (NO_2^-) is most often facilitated by *Nitrosomonas* spp. ("nitroso" referring to the formation of a nitroso functional group). Nitrite oxidation to nitrate (NO_3^-), though traditionally believed to be facilitated by *Nitrobacter* spp. (nitro referring the formation of a nitro functional group), is now known to be facilitated in the environment almost exclusively by *Nitrospira* spp.

Denitrification requires anoxic conditions to encourage the appropriate biological communities to form. It is facilitated by a wide diversity of bacteria. Sand filters, lagooning and reed beds can all be used to reduce nitrogen, but the activated sludge process (if designed well) can do the job the most easily. Since denitrification is the reduction of nitrate to dinitrogen (molecular nitrogen) gas, an electron donor is needed. This can be, depending on the waste water, organic matter (from feces), sulfide, or an added donor like methanol. The sludge in the anoxic tanks (denitrification tanks) must be mixed well (mixture of recirculated mixed liquor, return activated sludge [RAS], and raw influent) e.g. by using submersible mixers in order to achieve the desired denitrification.

Sometimes the conversion of toxic ammonia to nitrate alone is referred to as tertiary treatment.

Over time, different treatment configurations have evolved as denitrification has become more sophisticated. An initial scheme, the Ludzack-Ettinger Process, placed an anoxic treatment zone before the aeration tank and clarifier, using the return activated sludge (RAS) from the clarifier as a nitrate source. Influent wastewater (either raw or as effluent from primary clarification) serves as the electron source for the facultative bacteria to metabolize carbon, using the inorganic nitrate as a source of oxygen instead of dissolved molecular oxygen. This denitrification scheme was naturally limited to the amount of soluble nitrate present in the RAS. Nitrate reduction was limited because RAS rate is limited by the performance of the clarifier.

The "Modified Ludzak-Ettinger Process" (MLE) is an improvement on the original concept, for it recycles mixed liquor from the discharge end of the aeration tank to the head of the anoxic tank to provide a consistent source of soluble nitrate for the facultative bacteria. In this instance, raw wastewater continues to provide the electron source, and sub-surface mixing maintains the bacteria in contact with both electron source and soluble nitrate in the absence of dissolved oxygen.

Many sewage treatment plants use centrifugal pumps to transfer the nitrified mixed liquor from the aeration zone to the anoxic zone for denitrification. These pumps are often referred to as *Internal Mixed Liquor Recycle* (IMLR) pumps. IMLR may be 200% to 400% the flow rate of influent wastewater (Q.) This is in addition to Return Activated Sludge (RAS) from secondary clarifiers, which may be 100% of Q. (Therefore, the hydraulic capacity of the tanks in such a system should handle at least 400% of annual average design flow (AADF.) At times, the raw or primary effluent wastewater must be carbon-supplemented by the addition of methanol, acetate, or simple food waste (molasses, whey, plant starch) to improve the treatment efficiency. These carbon additions should be accounted for in the design of a treatment facility's organic loading.

Further modifications to the MLE were to come: Bardenpho and Biodenipho processes include additional anoxic and oxidative processes to further polish the conversion of nitrate ion to molecular nitrogen gas. Use of an anaerobic tank following the initial anoxic process allows for luxury uptake of phosphorus by bacteria, thereby biologically reducing orthophosphate ion in the treated wastewater. Even newer improvements, such as Anammox Process, interrupt the formation of nitrate at the nitrite stage of nitrification, shunting nitrite-rich mixed liquor activated sludge to treatment where nitrite is then converted to molecular nitrogen gas, saving energy, alkalinity, and secondary carbon sourcing. Anammox™ (ANaerobic AMMonia OXidation) works by artificially extending detention time and preserving denitrifiying bacteria through the use of substrate added to the mixed liquor and continuously recycled from it prior to secondary clarification. Many other proprietary schemes are being deployed, including DEMON™, Sharon-ANAMMOX™, ANITA-Mox™, and DeAmmon™. The bacteria Brocadia anammoxidans can remove ammonium from waste water through anaerobic oxidation of ammonium to hydrazine, a form of rocket fuel.

Phosphorus Removal

Every adult human excretes between 200 and 1000 grams of phosphorus annually. Studies of United States sewage in the late 1960s estimated mean per capita contributions of 500 grams in urine and feces, 1000 grams in synthetic detergents, and lesser variable amounts used as corrosion and scale control chemicals in water supplies. Source control via alternative detergent formulations has subsequently reduced the largest contribution, but the content of urine and feces will remain unchanged. Phosphorus removal is important as it is a limiting nutrient for algae growth in many fresh water systems. It is also particularly important for water reuse systems where high phosphorus concentrations may lead to fouling of downstream equipment such as reverse osmosis.

Phosphorus can be removed biologically in a process called enhanced biological phosphorus removal. In this process, specific bacteria, called polyphosphate-accumulating organisms (PAOs), are selectively enriched and accumulate large quantities of phosphorus within their cells (up to 20 percent of their mass). When the biomass enriched in these bacteria is separated from the treated water, these biosolids have a high fertilizer value.

Phosphorus removal can also be achieved by chemical precipitation, usually with salts of iron (e.g. ferric chloride), aluminum (e.g. alum), or lime. This may lead to excessive sludge production as hydroxides precipitates and the added chemicals can be expensive. Chemical phosphorus removal requires significantly smaller equipment footprint than biological removal, is easier to operate and is often more reliable than biological phosphorus removal. Another method for phosphorus removal is to use granular laterite.

Once removed, phosphorus, in the form of a phosphate-rich sewage sludge, may be dumped in a landfill or used as fertilizer. In the latter case, the treated sewage sludge is also sometimes referred to as biosolids.

Disinfection

The purpose of disinfection in the treatment of waste water is to substantially reduce the number of microorganisms in the water to be discharged back into the environment for the later use of drinking, bathing, irrigation, etc. The effectiveness of disinfection depends on the quality of the water being treated (e.g., cloudiness, pH, etc.), the type of disinfection being used, the disinfectant dosage (concentration and time), and other environmental variables. Cloudy water will be treated less successfully, since solid matter can shield organisms, especially from ultraviolet light or if contact times are low. Generally, short contact times, low doses and high flows all militate against effective disinfection. Common methods of disinfection include ozone, chlorine, ultraviolet light, or sodium hypochlorite. Chloramine, which is used for drinking water, is not used in the treatment of waste water because of its persistence. After multiple steps of disinfection, the treated water is ready to be released back into the water cycle by means of the nearest body of water or agriculture. Afterwards, the water can be transferred to reserves for everyday human uses.

Chlorination remains the most common form of waste water disinfection in North America due to its low cost and long-term history of effectiveness. One disadvantage is that chlorination of residual organic material can generate chlorinated-organic compounds that may be carcinogenic or harmful to the environment. Residual chlorine or chloramines may also be capable of chlorinating organic material in the natural aquatic environment. Further, because residual chlorine is toxic to aquatic species, the treated effluent must also be chemically dechlorinated, adding to the complexity and cost of treatment.

Ultraviolet (UV) light can be used instead of chlorine, iodine, or other chemicals. Because no chemicals are used, the treated water has no adverse effect on organisms that later consume it, as may be the case with other methods. UV radiation causes damage to the genetic structure of bacteria, viruses, and other pathogens, making them incapable of reproduction. The key disadvantages of UV disinfection are the need for frequent lamp maintenance and replacement and the need for a highly treated effluent to ensure that the target microorganisms are not shielded from the UV radiation (i.e., any solids present in the treated effluent may protect microorganisms from the UV light). In the United Kingdom, UV light is becoming the most common means of disinfection because of the concerns about the impacts of chlorine in chlorinating residual organics in the wastewater and in chlorinating organics in the receiving water. Some sewage treatment systems in Canada and the US also use UV light for their effluent water disinfection.

Ozone (O_3) is generated by passing oxygen (O_2) through a high voltage potential resulting in a third oxygen atom becoming attached and forming O_3. Ozone is very unstable and reactive and oxidizes most organic material it comes in contact with, thereby destroying many pathogenic microorganisms. Ozone is considered to be safer than chlorine because, unlike chlorine which has to be stored on site (highly poisonous in the event of an accidental release), ozone is generated on-site as needed. Ozonation also produces fewer disinfection by-products than chlorination. A disadvantage of ozone disinfection is the high cost of the ozone generation equipment and the requirements for special operators.

Fourth Treatment Stage

Micropollutants such as pharmaceuticals, ingredients of household chemicals, chemicals used in small businesses or industries, environmental persistent pharmaceutical pollutant (EPPP) or pesticides may not be eliminated in the conventional treatment process (primary, secondary and tertiary treatment) and therefore lead to water pollution. Although concentrations of those substances and their decompostion products are quite low, there is still a chance to harm aquatic organisms. For pharmaceuticals, the following substances have been identified as "toxicologically relevant": substances with endocrine disrupting effects, genotoxic substances and substances that enhance the development of bacterial resistances. They mainly belong to the group of environmental persistent pharmaceutical pollutants. Techniques for elimination of micropollutants via a fourth treatment stage during sewage treatment are being tested in Germany, Switzerland and the Netherlands. However, since those techniques are still costly, they are not yet applied on a regular basis. Such process steps mainly consist of activated carbon filters that adsorb the micropollutants. Ozone can also be applied as an oxidative method. Also the use of enzymes such as the enzyme laccase is under investigation. A new concept which could provide an energy-efficient treatment of micropollutants could be the use of laccase secreting fungi cultivated at a wastewater treatment plant to degrade micropollutants and at the same time to provide enzymes at a cathode of a microbial biofuel cells. Microbial biofuel cells are investigated for their property to treat organic matter in wastewater.

To reduce pharmaceuticals in water bodies, also "source control" measures are under investigation, such as innovations in drug development or more responsible handling of drugs.

Odor Control

Odors emitted by sewage treatment are typically an indication of an anaerobic or "septic" condition. Early stages of processing will tend to produce foul-smelling gases, with hydrogen sulfide being most common in generating complaints. Large process plants in urban areas will often treat the odors with carbon reactors, a contact media with bio-slimes, small doses of chlorine, or circulating fluids to biologically capture and metabolize the noxious gases. Other methods of odor control exist, including addition of iron salts, hydrogen peroxide, calcium nitrate, etc. to manage hydrogen sulfide levels.

High-density solids pumps are suitable for reducing odors by conveying sludge through hermetic closed pipework.

Energy Requirements

For conventional sewage treatment plants, around 30 percent of the annual operating costs is usually required for energy. The energy requirements vary with type of treatment process as well as wastewater load. For example, constructed wetlands have a lower energy requirement than activated sludge plants, as less energy is required for the aeration step. Sewage treatment plants that produce biogas in their sewage sludge treatment process with anaerobic digestion can produce enough energy to meet most of the energy needs of the sewage treatment plant itself.

In conventional secondary treatment processes, most of the electricity is used for aeration, pumping systems and equipment for the dewatering and drying of sewage sludge. Advanced wastewater treatment plants, e.g. for nutrient removal, require more energy than plants that only achieve primary or secondary treatment.

Sludge Treatment and Disposal

The sludges accumulated in a wastewater treatment process must be treated and disposed of in a safe and effective manner. The purpose of digestion is to reduce the amount of organic matter and the number of disease-causing microorganisms present in the solids. The most common treatment options include anaerobic digestion, aerobic digestion, and composting. Incineration is also used, albeit to a much lesser degree.

Sludge treatment depends on the amount of solids generated and other site-specific conditions. Composting is most often applied to small-scale plants with aerobic digestion for mid-sized operations, and anaerobic digestion for the larger-scale operations.

The sludge is sometimes passed through a so-called pre-thickener which de-waters the sludge. Types of pre-thickeners include centrifugal sludge thickeners rotary drum sludge thickeners and belt filter presses. Dewatered sludge may be incinerated or transported offsite for disposal in a landfill or use as an agricultural soil amendment.

Environment Aspects

The outlet of the Karlsruhe sewage treatment plant flows into the Alb

Many processes in a wastewater treatment plant are designed to mimic the natural treatment processes that occur in the environment, whether that environment is a natural water body or the ground. If not overloaded, bacteria in the environment will consume organic contaminants, although this will reduce the levels of oxygen in the water and may significantly change the overall ecology of the receiving water. Native bacterial populations feed on the organic contaminants, and the numbers of disease-causing microorganisms are reduced by natural environmental conditions such as predation or exposure to ultraviolet radiation. Consequently, in cases where the receiving environment provides a high level of dilution, a high degree of wastewater treatment may not be required. However, recent evidence has demonstrated that very low levels of specific contaminants in wastewater, including hormones (from animal husbandry and residue from human hormonal contraception methods) and synthetic materials such as phthalates that mimic hormones in their action, can have an unpredictable adverse impact on the natural biota and potentially on humans if the water is re-used for drinking water. In the US and EU, uncontrolled discharges of wastewater to the environment are not permitted under law, and strict water quality requirements are to be met, as clean drinking water is essential. A significant threat in the coming decades will be the increasing uncontrolled discharges of wastewater within rapidly developing countries.

Effects on Biology

Sewage treatment plants can have multiple effects on nutrient levels in the water that the treated sewage flows into. These nutrients can have large effects on the biological life in the water in contact with the effluent. Stabilization ponds (or sewage treatment ponds) can include any of the following:

- Oxidation ponds, which are aerobic bodies of water usually 1–2 meters in depth that receive effluent from sedimentation tanks or other forms of primary treatment.

  - Dominated by algae

- Polishing ponds are similar to oxidation ponds but receive effluent from an oxidation pond or from a plant with an extended mechanical treatment.

  - Dominated by zooplankton

- Facultative lagoons, raw sewage lagoons, or sewage lagoons are ponds where sewage is added with no primary treatment other than coarse screening. These ponds provide effective treatment when the surface remains aerobic; although anaerobic conditions may develop near the layer of settled sludge on the bottom of the pond.

- Anaerobic lagoons are heavily loaded ponds.

  - Dominated by bacteria

- Sludge lagoons are aerobic ponds, usually 2 to 5 meters in depth, that receive anaerobically digested primary sludge, or activated secondary sludge under water.

  - Upper layers are dominated by algae

Phosphorus limitation is a possible result from sewage treatment and results in flagellate-dominated plankton, particularly in summer and fall.

A phytoplankton study found high nutrient concentrations linked to sewage effluents. High nutrient concentration leads to high chlorophyll a concentrations, which is a proxy for primary production in marine environments. High primary production means high phytoplankton populations and most likely high zooplankton populations, because zooplankton feed on phytoplankton. However, effluent released into marine systems also leads to greater population instability.

The planktonic trends of high populations close to input of treated sewage is contrasted by the bacterial trend. In a study of *Aeromonas* spp. in increasing distance from a wastewater source, greater change in seasonal cycles was found the furthest from the effluent. This trend is so strong that the furthest location studied actually had an inversion of the *Aeromonas* spp. cycle in comparison to that of fecal coliforms. Since there is a main pattern in the cycles that occurred simultaneously at all stations it indicates seasonal factors (temperature, solar radiation, phytoplankton) control of the bacterial population. The effluent dominant species changes from *Aeromonas caviae* in winter to *Aeromonas sobria* in the spring and fall while the inflow dominant species is *Aeromonas caviae*, which is constant throughout the seasons.

Treated Sewage Reuse

With suitable technology, it is possible to reuse sewage effluent for drinking water, although this is usually only done in places with limited water supplies, such as Windhoek and Singapore.

In Israel, about 50 percent of agricultural water use (total use was 1 billion cubic metres in 2008) is provided through reclaimed sewer water. Future plans call for increased use of treated sewer water as well as more desalination plants.

Sewage Treatment in Developing Countries

Few reliable figures exist on the share of the wastewater collected in sewers that is being treated in the world. A global estimate by UNDP and UN-Habitat is that 90% of all wastewater generated is released into the environment untreated. In many developing countries the bulk of domestic and industrial wastewater is discharged without any treatment or after primary treatment only.

In Latin America about 15 percent of collected wastewater passes through treatment plants (with varying levels of actual treatment). In Venezuela, a below average country in South America with respect to wastewater treatment, 97 percent of the country's sewage is discharged raw into the environment. In Iran, a relatively developed Middle Eastern country, the majority of Tehran's population has totally untreated sewage injected to the city's groundwater. However, the construction of major parts of the sewage system, collection and treatment, in Tehran is almost complete, and under development, due to be fully completed by the end of 2012. In Isfahan, Iran's third largest city, sewage treatment was started more than 100 years ago.

Only few cities in sub-Saharan Africa have sewer-based sanitation systems, let alone wastewater treatment plants, an exception being South Africa and – until the late 1990s- Zimbabwe. Instead, most urban residents in sub-Saharan Africa rely on on-site sanitation systems without sewers, such as septic tanks and pit latrines, and faecal sludge management in these cities is an enormous challenge.

History

FARADAY GIVING HIS CARD TO FATHER THAMES;
And we hope the Dirty Fellow will consult the learned Professor.

The Great Stink of 1858 stimulated research into the problem of sewage treatment. In this caricature in *The Times*, Michael Faraday reports to *Father Thames* on the state of the river

Basic sewer systems were used for waste removal in ancient Mesopotamia, where vertical shafts carried the waste away into cesspools. Similar systems existed in the Indus Valley civilization in modern-day India and in Ancient Crete and Greece. In the Middle Ages the sewer systems built by the Romans fell into disuse and waste was collected into cesspools that were periodically emptied by workers known as 'rakers' who would often sell it as fertilizer to farmers outside the city.

Modern sewage systems were first built in the mid-nineteenth century as a reaction to the exacerbation of sanitary conditions brought on by heavy industrialization and urbanization. Due to the contaminated water supply, cholera outbreaks occurred in 1832,

1849 and 1855 in London, killing tens of thousands of people. This, combined with the Great Stink of 1858, when the smell of untreated human waste in the River Thames became overpowering, and the report into sanitation reform of the Royal Commissioner Edwin Chadwick, led to the Metropolitan Commission of Sewers appointing Sir Joseph Bazalgette to construct a vast underground sewage system for the safe removal of waste. Contrary to Chadwick's recommendations, Bazalgette's system, and others later built in Continental Europe, did not pump the sewage onto farm land for use as fertilizer; it was simply piped to a natural waterway away from population centres, and pumped back into the environment.

Early Attempts

One of the first attempts at diverting sewage for use as a fertilizer in the farm was made by the cotton mill owner James Smith in the 1840s. He experimented with a piped distribution system initially proposed by James Vetch that collected sewage from his factory and pumped it into the outlying farms, and his success was enthusiastically followed by Edwin Chadwick and supported by organic chemist Justus von Liebig.

The idea was officially adopted by the Health of Towns Commission, and various schemes (known as sewage farms) were trialled by different municipalities over the next 50 years. At first, the heavier solids were channeled into ditches on the side of the farm and were covered over when full, but soon flat-bottomed tanks were employed as reservoirs for the sewage; the earliest patent was taken out by William Higgs in 1846 for "tanks or reservoirs in which the contents of sewers and drains from cities, towns and villages are to be collected and the solid animal or vegetable matters therein contained, solidified and dried..." Improvements to the design of the tanks included the introduction of the horizontal-flow tank in the 1850s and the radial-flow tank in 1905. These tanks had to be manually de-sludged periodically, until the introduction of automatic mechanical de-sludgers in the early 1900s.

The precursor to the modern septic tank was the cesspool in which the water was sealed off to prevent contamination and the solid waste was slowly liquified due to anaerobic action; it was invented by L.H Mouras in France in the 1860s. Donald Cameron, as City Surveyor for Exeter patented an improved version in 1895, which he called a 'septic tank'; septic having the meaning of 'bacterial'. These are still in worldwide use, especially in rural areas unconnected to large-scale sewage systems.

Chemical Treatment

It was not until the late 19th century that it became possible to treat the sewage by chemically breaking it down through the use of microorganisms and removing the pollutants. Land treatment was also steadily becoming less feasible, as cities grew and the volume of sewage produced could no longer be absorbed by the farmland on the outskirts.

Sir Edward Frankland, a distinguished chemist, who demonstrated the possibility of chemically treating sewage in the 1870s

Sir Edward Frankland conducted experiments at the Sewage Farm in Croydon, England, during the 1870s and was able to demonstrate that filtration of sewage through porous gravel produced a nitrified effluent (the ammonia was converted into nitrate) and that the filter remained unclogged over long periods of time. This established the then revolutionary possibility of biological treatment of sewage using a contact bed to oxidize the waste. This concept was taken up by the chief chemist for the London Metropolitan Board of Works, William Libdin, in 1887:

> ...in all probability the true way of purifying sewage...will be first to separate the sludge, and then turn into neutral effluent... retain it for a sufficient period, during which time it should be fully aerated, and finally discharge it into the stream in a purified condition. This is indeed what is aimed at and imperfectly accomplished on a sewage farm.

From 1885 to 1891 filters working on this principle were constructed throughout the UK and the idea was also taken up in the US at the Lawrence Experiment Station in Massachusetts, where Frankland's work was confirmed. In 1890 the LES developed a 'trickling filter' that gave a much more reliable performance.

Contact beds were developed in Salford, Lancashire and by scientists working for the London City Council in the early 1890s. According to Christopher Hamlin, this was part of a conceptual revolution that replaced the philosophy that saw "sewage purification as the prevention of decomposition with one that tried to facilitate the biological process that destroy sewage naturally."

Contact beds were tanks containing the inert substance, such as stones or slate, that maximized the surface area available for the microbial growth to break down the sewage. The sewage was held in the tank until it was fully decomposed and it was then filtered out into the ground. This method quickly became widespread, especially in the UK, where it was used in Leicester, Sheffield, Manchester and Leeds. The bacterial bed was simultaneously developed by Joseph Corbett as Borough Engineer in Salford and experiments in 1905 showed that his method was superior in that greater volumes of sewage could be purified better for longer periods of time than could be achieved by the contact bed.

The Royal Commission on Sewage Disposal published its eighth report in 1912 that set what became the international standard for sewage discharge into rivers; the '20:30 standard', which allowed 20 mg Biochemical oxygen demand and 30 mg suspended solid per litre.

Aeration Basics

Factors Affecting Removal of compounds by Aeration

- Physico- chemical properties of compound to be removed like hydrophobicity, surface area, etc.

- Temperature of water & air.

- Process parameters for aeration like air to water ratio, available area of mass transfer, contact time, etc.

Calculation of Solubility of Gases

Henrys' Law is defined as:

$$p_A = Hx_A$$

Where p_A is the partial pressure of any compound A in air (atm), H is the Henrys' constant which depends upon temperature and x_A is the mol fraction of compound A in water.

By definition

$$x_A = \frac{\text{Moles of compound A in liquid solution}}{\text{Moles of compound A in liquid solution} + \text{Moles of water in liquid solution}}$$

Since moles of oxygen in liquid solution are usually very less as compared to moles of water in liquid solution, therefore,

$$x_A = \frac{\text{Moles of compound A in liquid solution}}{\text{Moles of water in liquid solution}} = \frac{\text{moles of compound A}}{\text{moles of water}}$$

$$x_A = \frac{\text{moles of compound A}}{\text{moles of water}} = \frac{\left(\text{Weight of compound A (in g)} / \text{Molecular Weight}\right)}{\left(\left(\text{Density} \times \text{Volume of Water}\right) / \text{Molecular Weight of Water}\right)}$$

$$x_A = \frac{\left(\text{Weight of compound A (in g)} / \text{Molecular Weight}\right)}{\left(\left(1000 / 18\right) \times \text{Volume of Water (in Litre)}\right)}$$

$$x_A = \left(\left(\text{Concentration of compound A (in g / L)} \times 18\right) / \left(\text{Molecular Weight of compound A (in g/mol)} \times 10^3\right)\right)$$

$$x_A = \left(\left(C_A \text{(in g / L)} \times 18\right) / \left(MW_A \text{(in g / mol)} \times 10^3\right)\right)$$

Putting in earlier equation

$$C_A(\text{in g}/\text{L}) = \frac{p_A \times \left(\text{MW}_A(\text{in g}/\text{mol}) \times 10^3\right)}{(H \times 18)}$$

$$C_A(\text{in mg}/\text{L}) = \frac{p_A \times \text{MW}_A(\text{in g}/\text{mol}) \times 10^6}{(H \times 18)}$$

Where, C_A is the solubility of compound A in water.

Variation of Solubility of Gases with Temperature

Solubility of gases decreases with an increase in temperature. The change in Henrys' constant with temperature can be computed using van't Hoff type of equation:

$$\log_{10} H = \frac{-\Delta H}{RT} + b$$

Where,- ΔH is the Heat of absorption in kcal/kmol, R is the gas constant (=1.987 kcal/K-kmol), T is temperature in K and b is a dimensionless empirical constant.

Problem 1: The particle pressure of O_2 in atmosphere is 0.21 atm. Find the concentration of O_2 in water (in mg O_2/ litre of water) at 20°C & 5°C. Given that for oxygen, Henry's constant (H) is equal to 4.3×10^4 atm at 20°C, $\Delta H = 1.45\ 10^3$ kcal/kmol, and b=7.11.

Solution: Given that: P_A = 0.21 atm, H = 4.3×10^4 atm at 20°C.

$$C_{O_2}(\text{in mg}/\text{L}) = \frac{p_{O_2} \times \text{MW}_{O_2}(\text{in g}/\text{mol}) \times 10^6}{(H \times 18)}$$

$$C_{O_2}(\text{in mg}/\text{L}) = \frac{0.21 \times 32 \times 10^6}{\left(4.3 \times 10^4 \times 18\right)} = 8.682\ \text{mg}\ O_2\ /\ \text{litre of water at 20°C}$$

Now calculating Henrys' constant at 5°C

$$\log_{5°C} = \frac{-1.45 \times 10^3}{1.987 \times (273)} + 7.11$$

$$H_{5°C} = 30650.64\ \text{atm}$$

$$C_{O_2}(\text{in mg}/\text{L}) = \frac{0.21 \times 32 \times 10^6}{(30650.64 \times 18)} = 12.17\ \text{mg}\ O_2\ /\ \text{litre of water at 5°C}$$

Aeration Types

- Diffused or Submerged Aeration: Submerged aeration systems are used in lakes, reservoirs, and wastewater treatment facilities to increase dissolved oxygen (DO) levels and promote water circulation. Submerged diffusers release air or pure oxygen bubbles at depth, producing a free, turbulent bubble-plume that rises to the water surface through buoyant forces. The ascending bubble plume entrains water, causing vertical circulation and lateral surface spreading. Oxygen transfers to the water across the bubble interfaces as the bubbles rise from the diffuser to the water surface.

- Spray aeration: Spray aeration removes low levels of volatile contaminants. In a spray aeration system, water enters through the top of the unit and emerges through spray heads in a fine mist. Treated water collects in a vented tank below the spray heads. Volatile contaminants are released and vented to the outside.

- Water fall type of aeration: It involves flow of water over media forming droplets or thin film of water so as to contact with air.

Removal of Vocs by Aeration in Complete Stirred Tank Reactor (Cstr)

VOC Removal by Surface Aeration in CSTR

A mass balance for VOC in CSTR having surface aeration is given as:

$$
\begin{pmatrix} \text{Rate of accumulation} \\ \text{of VOC within the} \\ \text{system boundary} \end{pmatrix} = \begin{pmatrix} \text{Rate of flow of} \\ \text{VOC into the} \\ \text{system boundary} \end{pmatrix} - \begin{pmatrix} \text{Rate of flow of} \\ \text{VOC out of the} \\ \text{system boundary} \end{pmatrix} + \begin{pmatrix} \text{Rate of VOC} \\ \text{removal by stripping} \\ \text{or surface aeration} \end{pmatrix}
$$

$$
V\frac{dC}{dt} = QC_{in} - QC + r_{VOC}V = QC_{in} - QC + \left[-\left(k_L a\right)_{VOC} \left(C - C_S\right) \right] V
$$

Where, V is the volume of the CSTR (m³), dC/dt is the rate of change of VOC in the CSTR, Q is the liquid flow rate in and out of the reactor (m³/s), C_{in} is the VOC concentration in influent to the CSTR (µg/m³) and C is the VOC concentration in effluent from the CSTR (µg/m³). RVOC is the rate of VOC mass transfer (µg/m³s) and is given as:

$$
r_{VOC} = -\left(k_L a\right)_{VOC} \left(C - C_S\right)
$$

Where, C_S is the saturation concentration of VOC in the liquid (µg/m³) and $\left(k_L a\right)_{VOC}$ is overall mass transfer coefficient (s⁻¹) and it is determined using the oxygen mass transfer coefficient, $\left(k_L a\right)_{O_2}$ using the following equation:

$$
\left(k_L a\right)_{VOC} = \left(k_L a\right)_{O_2} \left(\frac{D_{VOC}}{D_{O_2}}\right)^n
$$

Where, D_{VOC} and D_{O_2} are the diffusion coefficients of VOC and O_2 in water, respectively (cm²/s) and n is an empirical constant.

Assuming steady state condition, $V/Q=\tau$ and $C_s=0$, equation 1 becomes:

$$\frac{C_{in}-C}{\tau}=\left(k_L a\right)_{VOC} C$$

$$\text{Fraction of VOC removed}=1-\frac{C}{C_{in}}=1-\left[1+\left(k_L a\right)_{VOC} \tau\right]^{-1}$$

VOC Removal by Diffused Aeration in CSTR

At steady state, mass balance on VOC in diffused aeration system in CSTR is given as:

$$\begin{pmatrix} \text{Rate of in-flow of} \\ \text{VOC in the} \\ \text{liquid boundary} \end{pmatrix} = \begin{pmatrix} \text{Rate of out-flow} \\ \text{of VOC with the} \\ \text{liquid boundary} \end{pmatrix} + \begin{pmatrix} \text{Rate of out-flow} \\ \text{of VOC with} \\ \text{the exit gas} \end{pmatrix}$$

$$QC_{in} = QC+Q_g C_{g,e}$$

Where, Q_g is the diffused gas flow rate inside the CSTR (m³/s), $C_{g,e}$ is the VOC concentration in exit gas (μg/m³). Bielefeldt and Stensel gave the following relationship for $C_{g,e}$ with C:

$$C_{g,e} = H_u C\left[1-\exp\left\{-\phi\right\}\right]=H_u C\left[1-\exp\left\{\frac{-\left(k_L a\right)_{VOC} V}{H_u Q_g}\right\}\right]$$

Where, $H_u=(H/RT)$ is the dimensionless value of Henry's constant and ϕ is VOC saturation parameter. After putting the value of $C_{g,e}$ and rearranging, we get

$$\text{Fraction of VOC removed} = 1-\frac{C}{C_{in}}=1-\left[1+\frac{H_u Q_g}{Q}\exp\left\{\frac{-\left(k_L a\right)_{VOC} V}{H_u Q_g}\right\}\right]^{-1}$$

Problem: Wastewater flow rate in a complete-mix activated sludge reactor having volume=2000 m³ and depth=8 m is 6000 m³/d. If the influent concentration of benzene is 200 μg/m³ and that the air flow rate (at standard condition) is 100 m³/min, determine the fraction of benzene that can be stripped off if the complete mix activated sludge reactor is equipped with

(a)   surface aeration system

(b)   diffused-air aeration system

Also given that: Oxygen diffusivity=2.11 × 10⁻⁵ cm²/s, benzene diffusivity=0.96 × 10⁻⁵ cm²/s, temperature=20°C, n=1, Henry's constant=5.49×10⁻³ m³atm/mol.

Solution: First the value of $\left(k_L a\right)_{VOC}$ is determined.

$$\left(k_L a\right)_{VOC} = \left(k_L a\right)_{O_2} \left(\frac{D_{VOC}}{D_{O_2}}\right)^n$$

$$\left(k_L a\right)_{Benzene} = 6.2\,h^{-1} \times \left[\frac{0.96\times10^{-5}\ cm^2/s}{2.11\times10^{-5}\ cm^2/s}\right]^1 = 2.8208\,h^{-1} = 0.047\,min^{-1}$$

Case a: Surface aeration system

$$\tau = \left(V/Q\right) = \left(2000/6000\right) = 0.333\,d{=}7.992\,h$$

$$\text{Fraction of VOC removed} = 1 - \frac{C}{C_{in}} = 1 - \left[1 + \left(k_L a\right)_{VOC} \tau\right]^{-1}$$

$$\text{Fraction of VOC removed} = 1 - \left[1 + 2.8208 \times 7.992\right]^{-1} = 0.9575$$

Case b: Diffused aeration system

$$H_u = \frac{H}{RT} = \frac{5.49\times10^{-3}}{0.0821\times\left(273.15+20\right)} = 0.2282$$

We need gas flow rate at the actual condition i.e. at half of the tank depth (=4 m) and at 20°C. We know:

$$\frac{P_{st}Q_{st}}{T_{st}} = \frac{P_{ac}Q_{g,ac}}{T_{ac}}$$

$$Q_g = Q_{g,ac} = \left(\frac{P_{st}}{P_{ac}}\right)\frac{T_{ac}}{T_{st}}Q_{st} = \left(\frac{P_{st}}{\rho_{water}gh}\right)\frac{T_{ac}}{T_{st}}Q_{st} = \left(\frac{1.013\times10^5}{1000\times9.81\times\left(8/2\right)}\right)\frac{293}{273}\times100 = 277.06\ m^3/min$$

$$\phi = \frac{\left(K_L a\right)_{VOC} \times V}{H_u Q_g} = \frac{0.047\times2000}{0.2282\times277.06} = 1.48675$$

$$Q = 6000\,m^3/d = 4.1667\,m^3/min$$

$$\text{Fraction of VOC removed} = 1 - \frac{C}{C_{in}} = 1 - \left[1 + \frac{H_u Q_g}{Q}\exp\{-\phi\}\right]^{-1}$$

$$\text{Fraction of VOC removed} = 1 - \left[1 + \frac{277.06\times0.2282}{4.1667}\exp\left(-1.48675\right)\right]^{-1} = 0.7738$$

Packed Tower Aeration

In packed tower aeration (PTA), wastewater to be treated in sprayed on the top of a tower. The tower is about 3-10 m in height and is packed with various types of packing which provide high surface area to volume ratio. Air is pumped simultaneously counter-currently through the packing from the bottom and removes the VOC from wastewater which itself in trickling over the packing. Air along with the VOC gets removed from the top while treated water is collected at the bottom.

Design of Packed-tower Aeration Unit

The height of the tower can be calculated using the following equation:

$$Z = HTU \times NTU$$

Where, HTU is the height of transfer unit and NTU is the number of transfer units. HTU represents the rate of mass transfer for a particular type of packing. It determines the efficiency of mass transfer from liquid to gas phase. It is related to liquid loading rate and is given by:

$$HTU = \frac{L}{k_L a C_o}$$

Where, L is the ratio of superficial molar to mass liquid flow, $K_L a$ is the overall mass transfer coefficient; C_o is the molar density of VOC in water (kmol/m³)

$$\frac{R}{R-1} \ln \left[\frac{\left\{ \frac{C_{in}}{C_{out}} (R-1) \right\} + 1}{R} \right]$$

Where, $R = H_u G/L$ is called the stripping factor, H_u is Henrys' constant dimensionless, $G = (Q_G / A)$ is the superfacial gas flow rate (kmol/h m²), A is the cross–sectional area of packed bed (m²) and Q_G is the gas flow rate (kmol/h).

Coagulation and Flocculation

Coagulation has been defined as the addition of a positively charged ion such as Al^{3+}, Fe^{3+} or catalytic polyelectrolyte that results in particle destabilization and charge neutralization.

- The purpose of coagulation is removal of finely divided suspended solids and colloidal material from the waste liquid.

- These contaminants cannot be separated by sedimentation alone except by the use of reasonably long detention periods; truly colloidal particles cannot be removed by settling.

- If these suspended pollutants are organic, they can often be oxidized by biological means, as on trickling filter; biochemical oxidation, however, is slower for suspended matter than for dissolved organic contaminants. If the quantity of insoluble organic matter is large, bio-oxidation equipment must be increased in size to care for this added duty; it is usually more economical to remove the greater part of such matter by chemical coagulation instead of in a trickling filter or activated sludge tank.

Flocculation and Settling

Flocculation is the formation of clumps or flocs of suspended solids by agglomeration of smaller suspended particles.

- Most chemical precipitates do not possess the property of flocculation to any appreciable degree, but rather tend to form dense, compact, crystalline particles that settle rapidly.

- Precipitates of ferric hydroxide, aluminum hydroxide, silica, and certain other substances formed by chemical reaction of coagulant chemicals, however, have the property of forming large flocs of high surface area. As these flocs move through the liquid in a settling tank, they remove other suspended solids by adsorption or mechanical sweeping, and hence perform a better clarification than could be achieved by plain sedimentation alone.

- Flocculation is aided by mild agitation for a period of 20 to 60 minutes, to allow time for maximum floc formation and growth.

- The agitation should be gentle, in order not to break flocs already formed. Gentle air agitation has also been employed to promote floc growth.

- After the floc has formed and grown to its most effective size, the waste passes to a sedimentation chamber for solids removal. Floc formation and growth may be retarded or stopped by surface-active chemicals such as soaps and synthetic detergents.

Coagulation Fundamentals

Colloidal solutions that do not agglomerate naturally are called stable. This is due their large surface-to-volume ratio resulting from their very small size. In these small particles, molecular arrangements within crystals, loss of atoms due to abrasion of the surfaces, or other factors causes their surfaces to be charged.

The colloids contained in the water are negatively charged at $pH > pH_{iso}$ and positively at $pH < pH_{iso}$. These colloids are stable due to the repulsive forces between the negative charges. These colloids are destabilized by positively charged ions formed from the hydrolysis of coagulants. Destabilization of colloidal particles can be influenced by the double layer compression, adsorption and charge neutralization, entrapment in precipitates (sweep flocculation) and interparticle bridging.

1. Double layer compression: The negative colloid and its positively charged atmosphere produce an electrical potential across the diffuse layer. This is highest at the surface and drops off progressively with distance, approaching zero at the outside of the diffuse layer and is known as Zeta potential.

   - When a coagulant is added, it destabilizes the negatively charged particles. A cationic coagulant such as a metal salt reduces the zeta potential of the particles by adding positive charge.

   - Double layer compression involves adding salts to the system. As the ionic concentration increases, the double layer and the repulsion energy curves are compressed until there is no longer an energy barrier. Particle agglomeration occurs rapidly under these conditions.

   - The thickness of the double layer depends upon the concentration of ions in solution. A higher level of ions means more positive ions are available to neutralize the colloid. The result is a thinner double layer.

2. Adsorption and charge neutralization: Inorganic coagulants (such as alum, Ferrous sulphate) often work through charge neutralization. When these metal based coagulants are added to water, it dissociates and metal ions formed. Fe^{2+}/Al^{3+} are liberated, if ferrous salt/alum is used. Liberated Fe^{2+}/Al^{3+} and OH– ions react to form various monomeric and polymeric hydrolyzed species.

   The concentration of the hydrolyzed metal species depends on the metal concentration, and the solution pH. The percentage of Fe^{2+} and Al^{3+} hydrolytic products can be calculated from the following stability constants:

$$Fe^{2+} \quad + \quad H_2O \quad =Fe(OH)^+ \quad + \quad H^+ \quad\quad pK_1 = 9.5$$
$$Fe(OH)^+ \ + \quad H_2O \quad =Fe(OH)_2 \quad + \quad H^+ \quad\quad pK_2 = 11.07$$
$$Fe(OH)_2 \ + \quad H_2O \quad =Fe(OH)_3^- \quad + \quad H^+ \quad\quad pK_3 = 10.43$$

$$Al^{3+} \quad + \quad H_2O \quad =Al(OH)^{2+} \quad + \quad H^+ \quad\quad pK_1 = 49.5$$
$$Al(OH)^{2+} \ + \quad H_2O \quad =Al(OH)_2^+ \quad + \quad H^+ \quad\quad pK_2 = 5.6$$
$$Al(OH)_2^+ \ + \quad H_2O \quad =Al(OH)_3 \quad + \quad H^+ \quad\quad pK_3 = 6.7$$
$$Al(OH)_3 \ + \quad H_2O \quad =Al(OH)_4^- \quad + \quad H^+ \quad\quad pK_4 = 5.6$$

The speciation diagram of Fe^{2+} and Al^{3+} as drawn using above stability constants is presented in Equation above, respectively.

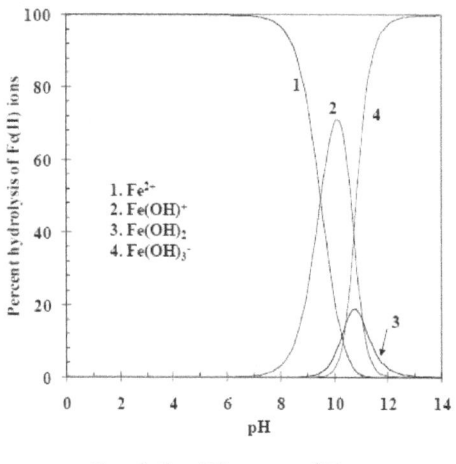

Speciation Diagram of Fe²⁺

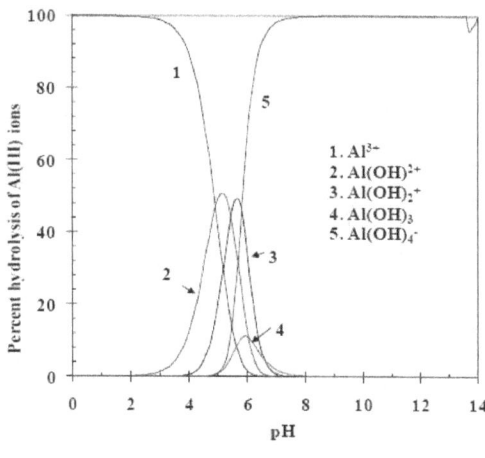

Speciation Diagram of Al³⁺

It can be seen from speciation diagram of Fe(II) ions that the dominant soluble species are Fe^{2+} and $Fe(OH)_3^-$ at low and high pH, respectively. The hydrolysis constants for aluminum cover a very narrower range, and all of the aluminum deprotonations are 'squeezed' into an interval of less than 2 unit. Therefore, apart from a narrow pH region approximately 5.5–6.5, the dominant soluble species are Al^{3+} and $Al(OH)_4^-$ at low and high pH, respectively.

Adsorption of the metal hydrolysed products on the colloid surface causes charge neutralization, which brings about van der Walls forces become dominant. Charge neutralization alone will not necessarily produce macro-flocs (flocs that can be seen with the naked eye). Micro-flocs (which are too small to be seen) may form but will not aggregate quickly into visible flocs. The polymeric hydrolyzed species possess high positive charges, and adsorbed to the surface of the negative colloids. This results in a reduction of the zeta potential to a level where the colloids are destabilized. The destabilized particles, along with their adsorbed hydro-metallic hydroxometallic complexes, aggregate by interparticulate Van der Waals forces. These forces are aided by the gentle mixing in water. When a coagulant forms threads or fibers which attach to several colloids, capturing and binding them together, this phenomenon is known as bridging. Some synthetic polymers and organic polyelectrolytes, instead of metallic salts, are used to assist interparticle bridging.

Adsorption sites on the colloidal particles can adsorb a polymer molecule. A bridge is formed when one or more particles become adsorbed along the length of the polymer. Bridge particles intertwined with other bridged particles during the process.

3. Sweep coagulation: Addition of relatively large doses of coagulants, usually aluminum or iron salts, which results in precipitation as hydrous metal oxides. Most of the colloids and some of dissolved solids are literally swept from the bulk of the water by becoming enmeshed in the settling hydrous oxide floc. This mechanism is often called sweep flocculation. Sweep floc is achieved by adding so much coagulant

to the water that the water becomes saturated and the coagulant precipitates out. Then the particles get trapped in the precipitant as it settles down.

Coagulation Reagents

Numerous chemicals are used in coagulation and flocculation processes. There are advantages and disadvantages associated with each chemical. Following factors should be considered in selecting these chemicals:

- Effectiveness.

- Cost.

- Reliability of supply.

- Sludge considerations.

- Compatibility with other treatment processes.

- Secondary pollution.

- Capital and operational costs for storage, feeding, and handling.

Coagulants and coagulant aids commonly used are generally classified as inorganic coagulants and polyelectrolytes. Polyelectrolytes are further classified as either synthetic-organic polymers or natural-organic polymers. The best choice is usually determined only after jar test is done in the laboratory.

Following table lists several common inorganic coagulants along with associated advantages and disadvantages.

Table: Advantages and disadvantages of alternative inorganic coagulants

Name	Advantages	Disadvantages
Aluminum Sulphate (Alum) $Al_2(SO_4)_3.18H_2O$	Easy to handle and apply; most commonly used; produces less sludge than lime; most effective between pH 6.5 and 7.5	Adds dissolved solids (salts) to water; effective over a limited pH range.
Sodium Aluminate $Na_2Al_2O_4$	Effective in hard waters; small dosages usually needed	Often used with alum; high cost; ineffective in soft waters
Polyaluminum Chloride (PAC) $Al_{13}(OH)_{20}(SO_4)_2.Cl_{15}$	In some applications, floc formed is more dense and faster settling than alum	Not commonly used; little full scale data compared to other aluminum derivatives
Ferric Sulphate $Fe_2(SO_4)_3$	Effective between pH 4–6 and 8.8–9.2	Adds dissolved solids (salts) to water; usually need to add alkalinity
Ferric Chloride $FeCl_3.6H_2O$	Effective between pH 4 and 11	Adds dissolved solids (salts) to water; consumes twice as much alkalinity as alum

Ferrous Sulphate (Copperas) $FeSO_4.7H_2O$	Not as pH sensitive as lime	Adds dissolved solids (salts) to water; usually need to add alkalinity
Lime $Ca(OH)_2$	Commonly used; very effective; may not add salts to effluent	Very pH dependent; produces large quantities of sludge; overdose can result in poor effluent quality

Polyelectrolytes

Polyelectrolytes are water-soluble polymers carrying ionic charge along the polymer chain and may be divided into natural and synthetic polyelectrolytes. Important natural polyelectrolytes include polymers of biological origin and those derived from starch products, cellulose derivatives and alginates. Depending on the type of charge, when placed in water, the polyelectrolytes are classified as anionic, cationic or nonionic.

- Anionic—ionize in solution to form negative sites along the polymer molecule.

- Cationic—ionize to form positive sites.

- Non-ionic—very slight ionization.

Common organic polyelectrolytes are shown in following table.

Table: Common organic polyelectrolytes

Polymer Type	Name	Mol. wt.	Available form	Typical use
Nonionic	Polyacrylamide	1×10^6 to 2×10^6	Powder,emulsion, solution	As flocculent with inorganic or organic polymers
Anionic	Hydrolyse Polyacryl-amide	1×10^6 to 2×10^7	Powder,emulsion, solution	As flocculent with inorganic or organic polymers
Cationic	Poly(DADMAC) or Poly(DADMAC) polymers	200 to 500×10^3	Solution	Primary coagulant alone or in combination with inorganics.
Cationic	Quaternized Poly-amines	10 to 500×10^4	Solution	Primary coagulant alone or in combination with inorganics.
Cationic	Polyamines	10^4 to 10^6	Solution	Primary coagulant alone or in combination with inorganics.

Polyelectrolytes Versus Inorganic Coagulants

Although they cannot be used exclusively, polyelectrolytes do possess several advantages over inorganic coagulants. These are as follows.

- During clarification, the volume of sludge produced can be reduced by 50 to 90%.

- The resulting sludge is more easily dewatered and contains less water.

- Polymeric coagulants do not affect pH. Therefore, the need for an alkaline chemical such as lime, caustic, or soda ash is reduced or eliminated.

- Polymeric coagulants do not add to the total dissolved solids concentration.

- Soluble iron or aluminum carryover in the clarifier effluent can result from inorganic coagulant use. By using polymeric coagulants, this problem can be reduced or eliminated.

Coagulant Aids

- In some waters, an even large dose of primary coagulant does not produce a satisfactory floc size and hence good settling rate. In these cases, a polymeric coagulant aid is added after the coagulant, to hasten reactions, to produce a denser floc, and thereby reducing the amount of primary coagulant required.

- Because of polymer bridging, small floc particles agglomerate rapidly into larger more cohesive floc, which settles rapidly.

- Coagulant aids also help to create satisfactory coagulation over a broader pH range.

- Generally, the most effective types of coagulant aids are slightly anionic polyacrylamides with very high-molecular weights.

- In some clarification systems, non-ionic or cationic types have proven effective.

Wastewater Treatment

Wastewater treatment plant in Cuxhaven, Germany

Wastewater treatment is a process used to convert wastewater - which is water no longer needed or suitable for its most recent use - into an effluent that can be either returned to the water cycle with minimal environmental issues or reused. The latter is called water reclamation and implies avoidance of disposal by use of treated wastewater effluent for various purposes. Treatment means removing impurities from water being treated; and some methods of treatment are applicable to both water and wastewater. The physical infrastructure used for wastewater treatment is called a wastewater treatment plant (WWTP).

The treatment of wastewater belongs to the overarching field of Public Works - Environmental, with the management of human waste, solid waste, sewage treatment, stormwater (drainage) management, and water treatment. By-products from wastewater treatment plants, such as screenings, grit and sewage sludge may also be treated in a wastewater treatment plant. If the wastewater is predominantly from municipal sources (households and small industries) it is called sewage and its treatment is called sewage treatment.

Disposal or Reuse

Although disposal or reuse occurs after treatment, it must be considered first. Since disposal or reuse are the objectives of wastewater treatment, disposal or reuse options are the basis for treatment decisions. Acceptable impurity concentrations may vary with the type of use or location of disposal. Transportation costs often make acceptable impurity concentrations dependent upon location of disposal, but expensive treatment requirements may encourage selection of a disposal location on the basis of impurity concentrations. Ocean disposal is subject to international treaty requirements. International treaties may also regulate disposal into rivers crossing international borders. Water bodies entirely within the jurisdiction of a single nation may be subject to regulations of multiple local governments. Acceptable impurity concentrations may vary widely among different jurisdictions for disposal of wastewater to evaporation ponds, infiltration basins, or injection wells.

Processes Used

Phase Separation

Clarifiers are widely used for wastewater treatment.

Phase separation transfers impurities into a non-aqueous phase. Phase separation may occur at intermediate points in a treatment sequence to remove solids generated during oxidation or polishing. Grease and oil may be recovered for fuel or saponification. Solids often require dewatering of sludge in a wastewater treatment plant. Disposal options for dried solids vary with the type and concentration of impurities removed from water.

Production of waste brine, however, may discourage wastewater treatment removing dissolved inorganic solids from water by methods like ion exchange, reverse osmosis, and distillation.

Primary settling tank of wastewater treatment plant in Dresden-Kaditz, Germany

Sedimentation

Solids and non-polar liquids may be removed from wastewater by gravity when density differences are sufficient to overcome dispersion by turbulence. Gravity separation of solids is the primary treatment of sewage, where the unit process is called "primary settling tanks" or "primary sedimentation tanks". It is also widely used for the treatment of other wastewaters. Solids that are heavier than water will accumulate at the bottom of quiescent settling basins. More complex clarifiers also have skimmers to simultaneously remove floating grease like soap scum and solids like feathers or wood chips. Containers like the API oil-water separator are specifically designed to separate non-polar liquids.

Filtration

Colloidal suspensions of fine solids may be removed by filtration through fine physical barriers distinguished from coarser screens or sieves by the ability to remove particles smaller than the openings through which the water passes. Other types of water filters remove impurities by chemical or biological processes described below.

Oxidation

Oxidation reduces the biochemical oxygen demand of wastewater, and may reduce the toxicity of some impurities. Secondary treatment converts some impurities to carbon dioxide, water, and biosolids. Chemical oxidation is widely used for disinfection.

Aeration tank of an activated sludge process at the wastewater treatment plant in Dresden-Kaditz, Germany

Biochemical Oxidation

Secondary treatment by biochemical oxidation of dissolved and colloidal organic compounds is widely used in sewage treatment and is applicable to some agricultural and industrial waste waters. Biological oxidation will preferentially remove organic compounds useful as a food supply for the treatment ecosystem. Concentration of some less digestible compounds may be reduced by co-metabolism. Removal efficiency is limited by the minimum food concentration required to sustain the treatment ecosystem.

Chemical Oxidation

Chemical oxidation may remove some persistent organic pollutants and concentrations remaining after biochemical oxidation. Disinfection by chemical oxidation kills bacteria and microbial pathogens by adding ozone, chlorine or hypochlorite to wastewater.

Polishing

Polishing refers to treatments made following the above methods. These treatments may also be used independently for some industrial wastewater. Chemical reduction or pH adjustment minimizes chemical reactivity of wastewater following chemical oxidation. Carbon filtering removes remaining contaminants and impurities by chemical absorption onto activated carbon. Filtration through sand (calcium carbonate) or fabric filters is the most common method used in municipal wastewater treatment.

Wastewater Treatment Plants

Wastewater treatment plants may be distinguished by the type of wastewater to be treated, i.e. whether it is sewage, industrial wastewater, agricultural wastewater or leachate.

Overview of the wastewater treatment plant of Antwerpen-Zuid, located in the south of the agglomeration of Antwerp (Belgium)

Sewage Treatment Plants

A typical municipal sewage treatment plant in an industrialized country may include primary treatment to remove solid material, secondary treatment to digest dissolved and suspended organic material as well as the nutrients nitrogen and phosphorus, and - sometimes but not always - disinfection to kill pathogenic bacteria. The sewage sludge that is produced in sewage treatment plants undergoes sludge treatment. Larger municipalities often include factories discharging industrial wastewater into the municipal sewer system. The term "sewage treatment plant" is now often replaced with the term "wastewater treatment plant".

Tertiary Treatment

Tertiary treatment is a term applied to polishing methods used following a traditional sewage treatment sequence. Tertiary treatment is being increasingly applied in industrialized countries and most common technologies are micro filtration or synthetic membranes. After membrane filtration, the treated wastewater is nearly indistinguishable from waters of natural origin of drinking quality (without its minerals). Nitrates can be removed from wastewater by natural processes in wetlands but also via microbial denitrification. Ozone wastewater treatment is also growing in popularity, and requires the use of an ozone generator, which decontaminates the water as ozone bubbles percolate through the tank, but this treatment is energy intensive. Latest, and very promising treatment technology is the use aerobic granulation.

Industrial Wastewater Treatment Plants

Disposal of wastewaters from an industrial plant is a difficult and costly problem. Most petroleum refineries, chemical and petrochemical plants have onsite facilities to treat their wastewaters so that the pollutant concentrations in the treated wastewater comply with the local and/or national regulations regarding disposal of wastewaters into community treatment plants or into rivers, lakes or oceans. Constructed wetlands are being used in an increasing number of cases as they provided high quality and productive on-site treatment. Other industrial processes that produce a lot of waste-waters such as paper and pulp production has created environmental concern, leading to development of processes to recycle water use within plants before they have to be cleaned and disposed.

Industrial wastewater treatment plants are required where municipal sewage treatment plants are unavailable or cannot adequately treat specific industrial wastewaters. Industrial wastewater plants may reduce raw water costs by converting selected wastewaters to reclaimed water used for different purposes. Industrial wastewater treatment plants may reduce wastewater treatment charges collected by municipal sewage treatment plants by pre-treating wastewaters to reduce concentrations of pollutants measured to determine user fees.

Although economies of scale may favor use of a large municipal sewage treatment plant for disposal of small volumes of industrial wastewater, industrial wastewater treatment and disposal may be less expensive than correctly apportioned costs for larger volumes of industrial wastewater not requiring the conventional sewage treatment sequence of a small municipal sewage treatment plant.

An industrial wastewater treatment plant may include one or more of the following rather than the conventional primary, secondary, and disinfection sequence of sewage treatment:

- An API oil-water separator, for removing separate phase oil from wastewater.

- A clarifier, for removing solids from wastewater.

- A roughing filter, to reduce the biochemical oxygen demand of wastewater.

- A carbon filtration plant, to remove toxic dissolved organic compounds from wastewater.

- An advanced electrodialysis reversal (EDR) system with ion exchange membranes.

Agricultural Wastewater Treatment Plants

Agricultural wastewater treatment for continuous confined animal operations like milk and egg production may be performed in plants using mechanized treatment units similar to those described under industrial wastewater; but where land is available for ponds, settling basins and facultative lagoons may have lower operational costs for seasonal use conditions from breeding or harvest cycles.

Leachate Treatment Plants

Leachate treatment plants are used to treat leachate from landfills. Treatment options include: biological treatment, mechanical treatment by ultrafiltration, treatment with active carbon filters and reverse osmosis using disc tube module technology.

Industrial Wastewater Treatment

Industrial wastewater treatment covers the mechanisms and processes used to treat wastewater that is produced as a by-product of industrial or commercial activities. After treatment, the treated industrial wastewater (or effluent) may be reused or released to a sanitary sewer or to a surface water in the environment. Most industries produce some wastewater although recent trends in the developed world have been to minimise such production or recycle such wastewater within the production process. However, many industries remain dependent on processes that produce wastewaters.

Sources of Industrial Wastewater

Complex Organic Chemicals Industry

A range of industries manufacture or use complex organic chemicals. These include pesticides, pharmaceuticals, paints and dyes, petrochemicals, detergents, plastics, paper pollution, etc. Waste waters can be contaminated by feedstock materials, by-products, product material in soluble or particulate form, washing and cleaning agents, solvents and added value products such as plasticisers. Treatment facilities that do not need control of their effluent typically opt for a type of aerobic treatment, i.e. aerated lagoons.

Electric Power Plants

Fossil-fuel power stations, particularly coal-fired plants, are a major source of industrial wastewater. Many of these plants discharge wastewater with significant levels of metals such as lead, mercury, cadmium and chromium, as well as arsenic, selenium and nitrogen compounds (nitrates and nitrites). Wastewater streams include flue-gas desulfurization, fly ash, bottom ash and flue gas mercury control. Plants with air pollution controls such as wet scrubbers typically transfer the captured pollutants to the wastewater stream.

Ash ponds, a type of surface impoundment, are a widely used treatment technology at coal-fired plants. These ponds use gravity to settle out large particulates (measured as total suspended solids) from power plant wastewater. This technology does not treat dissolved pollutants. Power stations use additional technologies to control pollutants, depending on the particular wastestream in the plant. These include dry ash handling, closed-loop ash recycling, chemical precipitation, biological treatment (such as an activated sludge process), membrane systems, and evaporation-crystallization systems. Technological advancements in ion exchange membranes and electrodialysis systems has enabled high efficiency treatment of flue-gas desulfurization wastewater to meet recent EPA discharge limits. The treatment approach is similar for other highly scaling industrial wastewaters.

Food Industry

Wastewater generated from agricultural and food operations has distinctive characteristics that set it apart from common municipal wastewater managed by public or private sewage treatment plants throughout the world: it is biodegradable and non-toxic, but has high concentrations of biochemical oxygen demand (BOD) and suspended solids (SS). The constituents of food and agriculture wastewater are often complex to predict, due to the differences in BOD and pH in effluents from vegetable, fruit, and meat products and due to the seasonal nature of food processing and post-harvesting.

Processing of food from raw materials requires large volumes of high grade water. Vegetable washing generates waters with high loads of particulate matter and some dissolved organic matter. It may also contain surfactants.

Animal slaughter and processing produces very strong organic waste from body fluids, such as blood, and gut contents. This wastewater is frequently contaminated by significant levels of antibiotics and growth hormones from the animals and by a variety of pesticides used to control external parasites.

Processing food for sale produces wastes generated from cooking which are often rich in plant organic material and may also contain salt, flavourings, colouring material and acids or alkali. Very significant quantities of oil or fats may also be present.

Iron and Steel Industry

The production of iron from its ores involves powerful reduction reactions in blast furnaces. Cooling waters are inevitably contaminated with products especially ammonia and cyanide. Production of coke from coal in coking plants also requires water cooling and the use of water in by-products separation. Contamination of waste streams includes gasification products such as benzene, naphthalene, anthracene, cyanide, ammonia, phenols, cresols together with a range of more complex organic compounds known collectively as polycyclic aromatic hydrocarbons (PAH).

The conversion of iron or steel into sheet, wire or rods requires hot and cold mechanical transformation stages frequently employing water as a lubricant and coolant. Contaminants include hydraulic oils, tallow and particulate solids. Final treatment of iron and steel products before onward sale into manufacturing includes *pickling* in strong mineral acid to remove rust and prepare the surface for tin or chromium plating or for other surface treatments such as galvanisation or painting. The two acids commonly used are hydrochloric acid and sulfuric acid. Wastewaters include acidic rinse waters together with waste acid. Although many plants operate acid recovery plants (particularly those using hydrochloric acid), where the mineral acid is boiled away from the iron salts, there remains a large volume of highly acid ferrous sulfate or ferrous chloride to be disposed of. Many steel industry wastewaters are contaminated by hydraulic oil, also known as *soluble oil*.

Mines and Quarries

Mine wastewater effluent in Peru, with neutralized pH from tailing runoff.

The principal waste-waters associated with mines and quarries are slurries of rock particles in water. These arise from rainfall washing exposed surfaces and haul roads and also from rock washing and grading processes. Volumes of water can be very high, especially rainfall related arisings on large sites. Some specialized separation operations, such as coal washing to separate coal from native rock using density gradients, can produce wastewater contaminated by fine particulate haematite and surfactants. Oils and hydraulic oils are also common contaminants.

Wastewater from metal mines and ore recovery plants are inevitably contaminated by the minerals present in the native rock formations. Following crushing and extraction of the desirable materials, undesirable materials may enter the wastewater stream. For metal mines, this can include unwanted metals such as zinc and other materials such as arsenic. Extraction of high value metals such as gold and silver may generate slimes containing very fine particles in where physical removal of contaminants becomes particularly difficult.

Additionally, the geologic formations that harbour economically valuable metals such as copper and gold very often consist of sulphide-type ores. The processing entails grinding the rock into fine particles and then extracting the desired metal(s), with the leftover rock being known as tailings. These tailings contain a combination of not only undesirable leftover metals, but also sulphide components which eventually form sulphuric acid upon the exposure to air and water that inevitably occurs when the tailings are disposed of in large impoundments. The resulting acid mine drainage, which is often rich in heavy metals (because acids dissolve metals), is one of the many environmental impacts of mining.

Nuclear Industry

The waste production from the nuclear and radio-chemicals industry is dealt with as *Radioactive waste*.

Pulp and Paper Industry

Effluent from the pulp and paper industry is generally high in suspended solids and BOD. Plants that bleach wood pulp for paper making may generate chloroform, dioxins (including 2,3,7,8-TCDD), furans, phenols and chemical oxygen demand (COD). Stand-alone paper mills using imported pulp may only require simple primary treatment, such as sedimentation or dissolved air flotation. Increased BOD or COD loadings, as well as organic pollutants, may require biological treatment such as activated sludge or upflow anaerobic sludge blanket reactors. For mills with high inorganic loadings like salt, tertiary treatments may be required, either general membrane treatments like ultrafiltration or reverse osmosis or treatments to remove specific contaminants, such as nutrients.

Industrial Oil Contamination

Industrial applications where oil enters the wastewater stream may include vehicle wash bays, workshops, fuel storage depots, transport hubs and power generation. Often the wastewater is discharged into local sewer or trade waste systems and must meet local environmental specifications. Typical contaminants can include solvents, detergents, grit. lubricants and hydrocarbons.

Water Treatment

Many industries have a need to treat water to obtain very high quality water for demanding purposes such as environmental discharge compliance. Water treatment produces organic and mineral sludges from filtration and sedimentation. Ion exchange using natural or synthetic resins removes calcium, magnesium and carbonate ions from water, typically replacing them with sodium, chloride, hydroxyl and/or other ions. Regeneration of ion exchange columns with strong acids and alkalis produces a wastewater rich in hardness ions which are readily precipitated out, especially when in admixture with other wastewater constituents.

Wool Processing

Insecticide residues in fleeces are a particular problem in treating waters generated in wool processing. Animal fats may be present in the wastewater, which if not contaminated, can be recovered for the production of tallow or further rendering.

Treatment of Industrial Wastewater

The various types of contamination of wastewater require a variety of strategies to remove the contamination.

Brine Treatment

Brine treatment involves removing dissolved salt ions from the waste stream. Although similarities to seawater or brackish water desalination exist, industrial brine treatment may contain unique combinations of dissolved ions, such as hardness ions or other metals, necessitating specific processes and equipment.

Brine treatment systems are typically optimized to either reduce the volume of the final discharge for more economic disposal (as disposal costs are often based on volume) or maximize the recovery of fresh water or salts. Brine treatment systems may also be optimized to reduce electricity consumption, chemical usage, or physical footprint.

Brine treatment is commonly encountered when treating cooling tower blowdown, produced water from steam assisted gravity drainage (SAGD), produced water from natural gas extraction such as coal seam gas, frac flowback water, acid mine or acid rock drainage, reverse osmosis reject, chlor-alkali wastewater, pulp and paper mill effluent, and waste streams from food and beverage processing.

Brine treatment technologies may include: membrane filtration processes, such as reverse osmosis; ion exchange processes such as electrodialysis or weak acid cation exchange; or evaporation processes, such as brine concentrators and crystallizers employing mechanical vapour recompression and steam.

Reverse osmosis may not be viable for brine treatment, due to the potential for fouling caused by hardness salts or organic contaminants, or damage to the reverse osmosis membranes from hydrocarbons.

Evaporation processes are the most widespread for brine treatment as they enable the highest degree of concentration, as high as solid salt. They also produce the highest purity effluent, even distillate-quality. Evaporation processes are also more tolerant of organics, hydrocarbons, or hardness salts. However, energy consumption is high and corrosion may be an issue as the prime mover is concentrated salt water. As a result, evaporation systems typically employ titanium or duplex stainless steel materials.

Brine Management

Brine management examines the broader context of brine treatment and may include consideration of government policy and regulations, corporate sustainability, environmental impact, recycling, handling and transport, containment, centralized compared to on-site treatment, avoidance and reduction, technologies, and economics. Brine management shares some issues with leachate management and more general waste management.

Solids Removal

Most solids can be removed using simple sedimentation techniques with the solids recovered as slurry or sludge. Very fine solids and solids with densities close to the density of water pose special problems. In such case filtration or ultrafiltration may be required. Although, flocculation may be used, using alum salts or the addition of poly-electrolytes.

Oils and Grease Removal

The effective removal of oils and grease is dependent on the characteristics of the oil in terms of its suspension state and droplet size, which will in turn affect the choice of separator technology.

Oil pollution in water usually comes in four states, often in combination:

- free oil - large oil droplets sitting on the surface;

- heavy oil, which sits at the bottom, often adhering to solids like dirt;

- emulsified, where the oil droplets are heavily "chopped"; and

- dissolved oil, where the droplets are fully dispersed and not visible. Emulsified oil droplets are the most common in industrial oily wastewater and are extremely difficult to separate.

The methodology for separating the oil is dependent on the oil droplet size. Larger oil droplets such as those in free oil pollution are easily removed, but as the droplets become smaller, some separator technologies perform better than others.

Most separator technologies will have an optimum range of oil droplet sizes that can be effectively treated. This is known as the "micron rating."

Analysing the oily water to determine droplet size can be performed with a video particle analyser. Alternatively, there are commonalities in industries for oil droplet sizes. Larger droplets–greater than 60 microns–are often present in wastewater in workshops, re-fuel areas and depots. Twenty to 50 micron oil droplets often are present in vehicle wash bays, meat processing and dairy manufacturing effluent and aluminium billet cooling towers. Smaller droplets in the range of 10 to 20 microns tend to occur in workshops and condensates.

Each separator technology will have its' own performance curve outlining optimum performance based on oil droplet size. the most common separators are gravity tanks or pits, API oil-water separators or plate packs, chemical treatment via DAFs, centrifuges, media filters and hydrocyclones.

API Separators

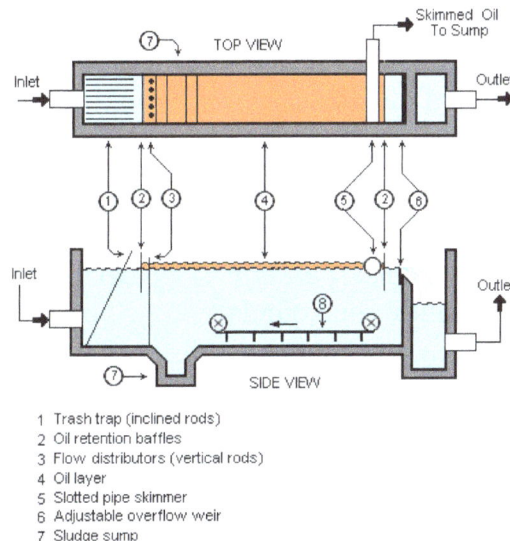

1 Trash trap (inclined rods)
2 Oil retention baffles
3 Flow distributors (vertical rods)
4 Oil layer
5 Slotted pipe skimmer
6 Adjustable overflow weir
7 Sludge sump
8 Chain and flight scraper

A typical API oil-water separator used in many industries

Many oils can be recovered from open water surfaces by skimming devices. Considered a dependable and cheap way to remove oil, grease and other hydrocarbons from water, oil skimmers can sometimes achieve the desired level of water purity. At other times, skimming is also a cost-efficient method to remove most of the oil before using membrane filters and chemical processes. Skimmers will prevent filters from blinding prematurely and keep chemical costs down because there is less oil to process.

Because grease skimming involves higher viscosity hydrocarbons, skimmers must be equipped with heaters powerful enough to keep grease fluid for discharge. If floating grease forms into solid clumps or mats, a spray bar, aerator or mechanical apparatus can be used to facilitate removal.

However, hydraulic oils and the majority of oils that have degraded to any extent will also have a soluble or emulsified component that will require further treatment to eliminate. Dissolving or emulsifying oil using surfactants or solvents usually exacerbates the problem rather than solving it, producing wastewater that is more difficult to treat.

The wastewaters from large-scale industries such as oil refineries, petrochemical plants, chemical plants, and natural gas processing plants commonly contain gross amounts of oil and suspended solids. Those industries use a device known as an API oil-water separator which is designed to separate the oil and suspended solids from their wastewater effluents. The name is derived from the fact that such separators are designed according to standards published by the American Petroleum Institute (API).

The API separator is a gravity separation device designed by using Stokes Law to define the rise velocity of oil droplets based on their density and size. The design is based on the specific gravity difference between the oil and the wastewater because that difference is much smaller than the specific gravity difference between the suspended solids and water. The suspended solids settles to the bottom of the separator as a sediment layer, the oil rises to top of the separator and the cleansed wastewater is the middle layer between the oil layer and the solids.

Typically, the oil layer is skimmed off and subsequently re-processed or disposed of, and the bottom sediment layer is removed by a chain and flight scraper (or similar device) and a sludge pump. The water layer is sent to further treatment for additional removal of any residual oil and then to some type of biological treatment unit for removal of undesirable dissolved chemical compounds.

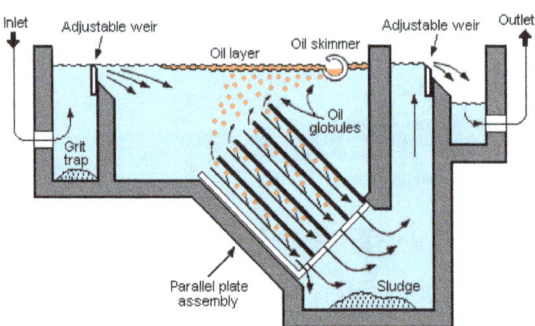

A typical parallel plate separator

Parallel plate separators arc similar to API separators but they include tilted parallel plate assemblies (also known as parallel packs). The parallel plates provide more surface for suspended oil droplets to coalesce into larger globules. Such separators still depend upon the specific gravity between the suspended oil and the water. However, the parallel plates enhance the degree of oil-water separation. The result is that a parallel plate separator requires significantly less space than a conventional API separator to achieve the same degree of separation.

Hydrocyclone Oil Separators

Hydrocyclone oil separators operate on the process where wastewater enters the cyclone chamber and is spun under extreme centrifugal forces more than 1000 times the force of gravity. This force causes the water and oil droplets to separate. The separated oil is discharged from one end of the cyclone where treated water is discharged through the opposite end for further treatment, filtration or discharge.

Hydrocyclones are useful for the greatest range of oil droplet sizes operating from less than 10 microns and up and can operate continuously without water pre-treatment and at any temperature and pH. Applications where hydrocyclones are found are in industry where oily water sources arise in workshops, vehicle wash bays, transport hubs, fuel depots and aluminium billet processing. Animal fats from meat processing and dairy manufacturing can also be removed without the need of chemical treatment that often is required for dissolved air flotation (DAF) systems.

Removal of Biodegradable Organics

Biodegradable organic material of plant or animal origin is usually possible to treat using extended conventional sewage treatment processes such as activated sludge or trickling filter. Problems can arise if the wastewater is excessively diluted with washing water or is highly concentrated such as undiluted blood or milk. The presence of cleaning agents, disinfectants, pesticides, or antibiotics can have detrimental impacts on treatment processes.

Activated Sludge Process

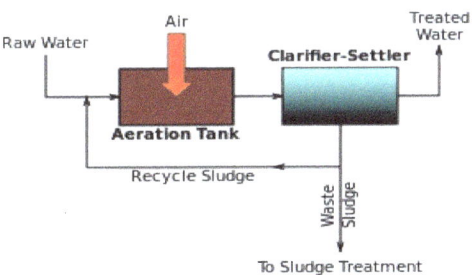

A generalized diagram of an activated sludge process.

Activated sludge is a biochemical process for treating sewage and industrial wastewater that uses air (or oxygen) and microorganisms to biologically oxidize organic pollutants, producing a waste sludge (or floc) containing the oxidized material. In general, an activated sludge process includes:

- An aeration tank where air (or oxygen) is injected and thoroughly mixed into the wastewater.

- A settling tank (usually referred to as a clarifier or "settler") to allow the waste sludge to settle. Part of the waste sludge is recycled to the aeration tank and the remaining waste sludge is removed for further treatment and ultimate disposal.

Trickling Filter Process

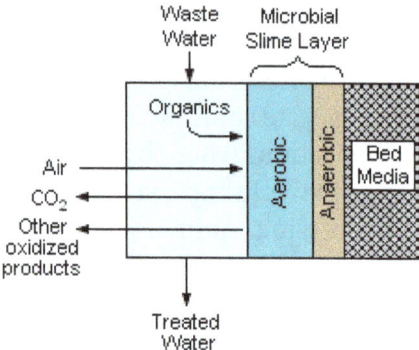

Image 1: A schematic cross-section of the contact face of the bed media in a trickling filter

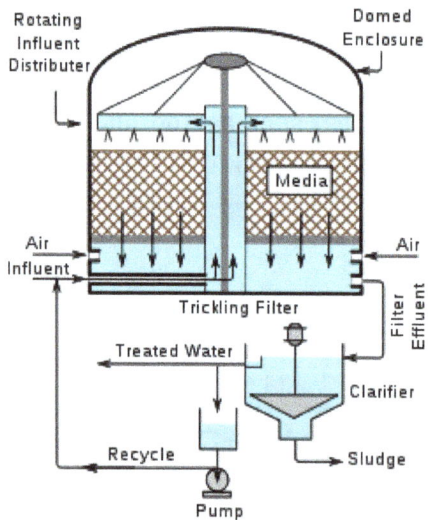

A typical complete trickling filter system

A trickling filter consists of a bed of rocks, gravel, slag, peat moss, or plastic media over which wastewater flows downward and contacts a layer (or film) of microbial slime covering the bed media. Aerobic conditions are maintained by forced air flowing through the bed or by natural convection of air. The process involves adsorption of organic compounds in the wastewater by the microbial slime layer, diffusion of air into the slime layer to provide the oxygen required for the biochemical oxidation of the organic compounds. The end products include carbon dioxide gas, water and other products of the oxidation. As the slime layer thickens, it becomes difficult for the air to penetrate the layer and an inner anaerobic layer is formed.

The fundamental components of a complete trickling filter system are:

- A bed of filter medium upon which a layer of microbial slime is promoted and developed.

- An enclosure or a container which houses the bed of filter medium.

- A system for distributing the flow of wastewater over the filter medium.

- A system for removing and disposing of any sludge from the treated effluent.

The treatment of sewage or other wastewater with trickling filters is among the oldest and most well characterized treatment technologies.

A trickling filter is also often called a *trickle filter*, *trickling biofilter*, *biofilter*, *biological filter* or *biological trickling filter*.

Removal of other Organics

Synthetic organic materials including solvents, paints, pharmaceuticals, pesticides, products from coke production and so forth can be very difficult to treat. Treatment methods are often specific to the material being treated. Methods include advanced oxidation processing, distillation, adsorption, vitrification, incineration, chemical immobilisation or landfill disposal. Some materials such as some detergents may be capable of biological degradation and in such cases, a modified form of wastewater treatment can be used.

Removal of Acids and Alkalis

Acids and alkalis can usually be neutralised under controlled conditions. Neutralisation frequently produces a precipitate that will require treatment as a solid residue that may also be toxic. In some cases, gases may be evolved requiring treatment for the gas stream. Some other forms of treatment are usually required following neutralisation.

Waste streams rich in hardness ions as from de-ionisation processes can readily lose the hardness ions in a buildup of precipitated calcium and magnesium salts. This precipitation process can cause severe *furring* of pipes and can, in extreme cases, cause the blockage of disposal pipes. A 1 metre diameter industrial marine discharge pipe serving a major chemicals complex was blocked by such salts in the 1970s. Treatment is by concentration of de-ionisation waste waters and disposal to landfill or by careful pH management of the released wastewater.

Removal of Toxic Materials

Toxic materials including many organic materials, metals (such as zinc, silver, cadmium, thallium, etc.) acids, alkalis, non-metallic elements (such as arsenic or selenium) are generally resistant to biological processes unless very dilute. Metals can often be precipitated out by changing the pH or by treatment with other chemicals. Many, however, are resistant to treatment or mitigation and may require concentration followed by landfilling or recycling. Dissolved organics can be *incinerated* within the wastewater by the advanced oxidation process.

Smart Capsules

Molecular encapsulation is a technology that has the potential to provide a system for the recyclable removal of lead and other ions from polluted sources. Nano-, micro- and milli- capsules, with sizes in the ranges 10 nm-1μm,1μm-1mm and >1mm, respectively, are particles that have an active reagent (core) surrounded by a carrier (shell).There are three types of capsule under investigation: alginate-based capsules, carbon nanotubes, polymer swelling capsules. These capsules provide a possible means for the remediation of contaminated water.

Agricultural Wastewater Treatment

Agricultural wastewater treatment is a farm management agenda for controlling pollution from surface runoff that may be contaminated by chemicals in fertiliser, pesticides, animal slurry, crop residues or irrigation water.

Nonpoint Source Pollution

Nonpoint source pollution from farms is caused by surface runoff from fields during rain storms. Agricultural runoff is a major source of pollution, in some cases the only source, in many watersheds.

Sediment Runoff

Highly erodible soils on a farm in Iowa

Soil washed off fields is the largest source of agricultural pollution in the United States. Excess sediment causes high levels of turbidity in water bodies, which can inhibit growth of aquatic plants, clog fish gills and smother animal larvae.

Farmers may utilize erosion controls to reduce runoff flows and retain soil on their fields. Common techniques include:

- contour ploughing

- crop mulching

- crop rotation

- planting perennial crops

- installing riparian buffers.

Nutrient Runoff

Manure spreader

Nitrogen and phosphorus are key pollutants found in runoff, and they are applied to farmland in several ways, such as in the form of commercial fertilizer, animal manure, or municipal or industrial wastewater (effluent) or sludge. These chemicals may also enter runoff from crop residues, irrigation water, wildlife, and atmospheric deposition.

Farmers can develop and implement nutrient management plans to mitigate impacts on water quality by:

- mapping and documenting fields, crop types, soil types, water bodies

- developing realistic crop yield projections

- conducting soil tests and nutrient analyses of manures and/or sludges applied

- identifying other significant nutrient sources (e.g., irrigation water)

- evaluating significant field features such as highly erodible soils, subsurface drains, and shallow aquifers

- applying fertilizers, manures, and/or sludges based on realistic yield goals and using precision agriculture techniques.

Pesticides

Aerial application (crop dusting) of pesticides over a soybean field in the U.S.

Pesticides are widely used by farmers to control plant pests and enhance production, but chemical pesticides can also cause water quality problems. Pesticides may appear in surface water due to:

- direct application (e.g. aerial spraying or broadcasting over water bodies)

- runoff during rain storms

- aerial drift (from adjacent fields).

Some pesticides have also been detected in groundwater.

Farmers may use Integrated Pest Management (IPM) techniques (which can include biological pest control) to maintain control over pests, reduce reliance on chemical pesticides, and protect water quality.

There are few safe ways of disposing of pesticide surpluses other than through containment in well managed landfills or by incineration. In some parts of the world, spraying on land is a permitted method of disposal.

Point Source Pollution

Farms with large livestock and poultry operations, such as factory farms, can be a major source of point source wastewater. In the United States, these facilities are called *concentrated animal feeding operations* or *confined animal feeding operations* and are being subject to increasing government regulation.

Animal Wastes

Confined Animal Feeding Operation in the United States

The constituents of animal wastewater typically contain

- Strong organic content — much stronger than human sewage

- High solids concentration

- High nitrate and phosphorus content

- Antibiotics

- Synthetic hormones

- Often high concentrations of parasites and their eggs

- Spores of *Cryptosporidium* (a protozoan) resistant to drinking water treatment processes

- Spores of *Giardia*

- Human pathogenic bacteria such as *Brucella* and *Salmonella*

Animal wastes from cattle can be produced as solid or semisolid manure or as a liquid slurry. The production of slurry is especially common in housed dairy cattle.

Treatment

Whilst solid manure heaps outdoors can give rise to polluting wastewaters from runoff, this type of waste is usually relatively easy to treat by containment and/or covering of the heap.

Animal slurries require special handling and are usually treated by containment in lagoons before disposal by spray or trickle application to grassland. Constructed wetlands are sometimes used to facilitate treatment of animal wastes, as are anaerobic lagoons. Excessive application or application to sodden land or insufficient land area can result in direct runoff to watercourses, with the potential for causing severe pollution. Application of slurries to land overlying aquifers can result in direct contamination or, more commonly, elevation of nitrogen levels as nitrite or nitrate.

The disposal of any wastewater containing animal waste upstream of a drinking water intake can pose serious health problems to those drinking the water because of the highly resistant spores present in many animals that are capable of causing disabling disease in humans. This risk exists even for very low-level seepage via shallow surface drains or from rainfall run-off.

Some animal slurries are treated by mixing with straws and composted at high temperature to produce a bacteriologically sterile and friable manure for soil improvement.

Piggery Waste

Hog confinement barn or piggery

Piggery waste is comparable to other animal wastes and is processed as for general animal waste, except that many piggery wastes contain elevated levels of copper that can be toxic in the natural environment. The liquid fraction of the waste is frequently separated off and re-used in the piggery to avoid the prohibitively expensive costs of disposing of copper-rich liquid. Ascarid worms and their eggs are also common in piggery waste and can infect humans if wastewater treatment is ineffective.

Silage Liquor

Fresh or wilted grass or other green crops can be made into a semi-fermented product called silage which can be stored and used as winter forage for cattle and sheep. The production of silage often involves the use of an acid conditioner such as sulfuric acid or formic acid. The process of silage making frequently produces a yellow-brown strongly smelling liquid which is very rich in simple sugars, alcohol, short-chain organic acids and silage conditioner. This liquor is one of the most polluting organic substances known. The volume of silage liquor produced is generally in proportion to the moisture content of the ensiled material.

Treatment

Silage liquor is best treated through prevention by wilting crops well before silage making. Any silage liquor that is produced can be used as part of the food for pigs. The most effective treatment is by containment in a slurry lagoon and by subsequent spreading on land following substantial dilution with slurry. Containment of silage liquor on its own can cause structural problems in concrete pits because of the acidic nature of silage liquor.

Milking Parlour (Dairy Farming) Wastes

Although milk has a deserved reputation as an important and valuable food product, its presence in wastewaters is highly polluting because of its organic strength, which can lead to very rapid de-oxygenation of receiving waters. Milking parlour wastes also contain large volumes of wash-down water, some animal waste together with cleaning and disinfection chemicals.

Treatment

Milking parlour wastes are often treated in admixture with human sewage in a local sewage treatment plant. This ensures that disinfectants and cleaning agents are sufficiently diluted and amenable to treatment. Running milking wastewaters into a farm slurry lagoon is a possible option although this tends to consume lagoon capacity very quickly. Land spreading is also a treatment option.

Slaughtering Waste

Wastewater from slaughtering activities is similar to milking parlour waste although considerably stronger in its organic composition and therefore potentially much more polluting.

Vegetable Washing Water

Washing of vegetables produces large volumes of water contaminated by soil and vegetable pieces. Low levels of pesticides used to treat the vegetables may also be present together with moderate levels of disinfectants such as chlorine.

Treatment

Most vegetable washing waters are extensively recycled with the solids removed by settlement and filtration. The recovered soil can be returned to the land.

Firewater

Although few farms plan for fires, fires are nevertheless more common on farms than on many other industrial premises. Stores of pesticides, herbicides, fuel oil for farm machinery and fertilizers can all help promote fire and can all be present in environmentally lethal quantities in firewater from fire fighting at farms.

Treatment

All farm environmental management plans should allow for containment of substantial quantities of firewater and for its subsequent recovery and disposal by specialist disposal companies. The concentration and mixture of contaminants in firewater make them unsuited to any treatment method available on the farm. Even land spreading has produced severe taste and odour problems for downstream water supply companies in the past.

Filtration

Water filtration is a mechanical or physical process of separating suspended and colloidal particles from fluids (liquids or gases) by interposing a medium through which only the fluid can pass. Medium used is generally a granular material through which water is passed. In the conventional water treatment process, filtration usually follows coagulation, flocculation, and sedimentation.

Filtration Process

- During filtration in a conventional down-flow depth filter, wastewater containing suspended matter is applied to the top of the filter bed.

- As the water passes through the filter bed, the suspended matter in the wastewater is removed by a variety of removal mechanisms.

- With passage of time, as material accumulates within the interstices of the granular medium, the head-loss through the filter starts to build up beyond the initial value.

- After some period of time, the operating head-loss or effluent turbidity reaches a predetermined head loss or turbidity value, and the filter must be cleaned (backwashed) to remove the material (suspended solids) that has accumulated within the granular filter bed.

- Backwashing is accomplished by reversing the flow through the filter. A sufficient flow of wash water is applied until the granular filtering medium is fluidized (expanded), causing the particles of the filtering medium to abrade against each other.

Filtration is classified into following three types :

1) Depth filtration

   a) Slow sand filtration

   b) Rapid porous and compressible medium filtration

   c) Intermittent porous medium filtration

   d) Recirculating porous medium filtration

2) Surface filtration

   a) Laboratory filters used for TSS test

   b) Diatomaceous earth filtration

   c) Cloth or screen filtration

3) Membrane flirtation

Depth Filtration

In this method, the removal of suspended particulate material from liquid slurry is done by passing the liquid through a filter bed composed of granular or compressible filter medium.

- Depth filtration is the solid/liquid separation process in which a dilute suspension or wastewater is passed through a packed bed of sand, anthracite, or other granular media.

- Solids (particles) get attached to the media or to the previously retained particles and are removed from the fluid.

- This method is virtually used everywhere in the treatment of surface waters for potable water supply.

- Depth filtration is also often successfully used as a tertiary treatment for wastewater.

- Failure of depth filtration affects the other downstream processes significantly and most of the times results in overall plant failure.

- Performance of a filter is quantified by particle removal efficiency and head loss across the packed bed.

- The duration of a filter run is limited by numerous constraints: available head, effluent quality or flow requirement.

- The head loss and removal efficiency of a filter are complicated functions of suspension qualities (particle size distribution and concentration, particle surface chemistry, and solution chemistry), filter design parameters (media size, type, and depth), and operating conditions (filtration rate and filter run-time).

Slow Sand Filtration (SSF):

- It is very effective for removing flocs containing microorganisms such as algae, bacteria, virus, etc.

- Slow sand filtration (SSF), with flow rates ranging between 0.1 and 0.2 m3 h−1, has been a standard biofiltration treatment for decades in the wastewater industry.

Rapid Sand Filtration (RSF)

- The major difference between SSF and RSF is in the principle of operation; that is, in the speed or rate at which water passes through the media.

- In Rapid sand filtration (RSF), water passes downward through a sand bed that removes the suspended particles.

- RSF is used today as an effective pretreatment procedure to enhance water quality prior to reverse osmosis (RO) membranes in desalination plants.

Surface Filtration

- Surface filtration involves removal of suspended material in a liquid by mechanical sieving. In this method, the liquid is passed through a thin septum (i.e., filter material).

- Materials that have been used as filter septum include woven metal fabrics, cloth fabrics of different weaves, and a variety of synthetic materials.

Membrane Filtration

- Membrane filtration can be broadly defined as a separation process that uses semi- permeable membrane to divide the feed stream into two portions: a permeate that contains the material passing through the membranes, and a retentate consisting of the species being left behind.

- Membrane filtration can be further classified in terms of the size range of permeating species, the mechanisms of rejection, the driving forces employed, the chemical structure and composition of membranes, and the geometry of construction.

- The most important types of membrane filtration are pressure driven processes including microfiltration (MF), ultrafiltration (UF), nanofiltration (NF), and reverse osmosis (RO).

Mechanisms Involved in the Filtration Processes

The process of filtration involves several mechanisms listed in the table. Straining has been identified as the principal mechanism that is operative in the removal of suspended solids during the filtration of settled secondary effluent from biological treatment processes. Other mechanisms including impaction, interception, and adhesion are also operative even though their effects are small and, for the most part, masked by the straining action.

Table: Mechanisms involved in the filtration processes

Mechanism/ phenomenon	Description
Straining a) Mechanical b) Chance contact	Particles larger than the pore space of the filtering medium are strained out mechanically. Particles smaller than the pore space are trapped within the filter by chance contact
Sedimentation	Particles settle on the filtering medium within the filter
Impaction	Heavy particles do not follow the flow streamlines
Interception	Particles get removed during contact with the surface of the filtering medium
Adhesion	Particles become attached to the surface of the filtering medium as they pass through.
Flocculation	It can occur within the interstices of the filter medium.
Chemical adsorption a) Bonding b) Chemical interaction	Once a particle has been brought in contact with the surface of the filtering medium or with other particles, either one of these mechanisms, chemical or physical adsorption or both, may occur.
Physical adsorption a) Electrostatic forces b) Electrokinetic forces c) Van der Waals forces	
Biological growth	Biological growth within the filter reduces the pore volume and enhances the removal of particles with any of the above removal mechanisms

Filter-Medium Characteristics

Grain size is the principle filter-medium characteristic that affects the filtration operation. Grain size affects both the clear-water head loss and the buildup of head loss during the filter run. If too small a filtering medium is selected, much of the driving force will be wasted in overcoming the frictional resistance of the filter bed. On the other hand, if the size of the medium is too large, many of the small particles in the influent will pass directly through the bed. The size distribution of the filter material is usually determined by sieve analysis using a series of decreasing sieve sizes.

Classification of Filters

Filters that must be taken off-line periodically to be backwashed are classified operationally as semi-continuous.

Filters in which is filtration and backwash operations occur simultaneously are classified as continuous.

Within each of these two classifications, there are a number of different types of filters depending on bed depth (e.g., shallow, conventional, and deep bed), the type filtering medium used (mono-, dual-, and multi-medium), whether the filtering medium is stratified or unstratified, the type of operation (down-flow or upflow), and the method used for the management of solids (surface or internal storage). For the mono- and dual-medium semi-continuous filters, a further classification can be made based on the driving force (e.g., gravity or pressure).

Types of Depth Filters

The five types of depth filters used most commonly for wastewater filtration are

(a) Conventional down-flow filters: Single-, dual-, or multimedium filter materials are utilized in conventional down-flow depth filters. Typically sand or anthracite is used as the filtering material in single-medium filters. Dual-medium filters usually consist of a layer anthracite over a layer of sand. Dual- and multimedium and deep-bed mono-medium depth filters were developed to allow the suspended solids in the liquid to be filtered to penetrate farther into the filter bed, and thus use more of the solids-storage capacity available within the filter bed.

(b) Deep-bed down-flow filters: The deep-bed down-flow filter is similar to the conventional down-flow filter with the exception that the depth of the filter bed and the size of the filter medium are greater than corresponding values an conventional filter. Because of the greater depth and larger medium size, more solids can be stored within the filter bed and the run length can be extended.

(c) Deep-bed upflow continuous-backwash filters: In this filter the wastewater to be filtered is introduced into the bottom of the filter where it flows upward through a series of riser tubes and is distributed evenly into the sand bed through the open bottom of an inlet distribution hood. The water then flows upward through the downward-moving sand. The clean filtrate exits from the sand bed, overflows a weir, and is discharged from the filter. Because the sand has higher settling velocity than the removed solids, the sand is not carried out of the filter.

(d) Pulsed-bed filter: The pulsed-bed filter is a proprietary down-flow gravity filter with an unstratified shallow layer of fine sand as the filtering medium. The shallow bed is used for solids storage, as opposed to other shallow-bed filters where solids are principally stored on the sand surface. An unusual feature of this filter is the use of an air pulse to disrupt the sand surface and thus allow penetration of suspended solids into the bed.

(e) Travelling-bridge filters: The travelling-bridge filter is a proprietary continuous down-flow, automatic backwash, low-head, granular medium depth filter. The bed of the filter is divided horizontally into long independent filter cells. Each filter cell contains approximately 280 mm of medium. Treated wastewater flows through the medium by gravity.

Secondary Treatment

Secondary treatment is a treatment process for wastewater (or sewage) to achieve a certain degree of effluent quality by using a sewage treatment plant with physical phase separation to remove settleable solids and a biological process to remove dissolved and suspended organic compounds. After this kind of treatment, the wastewater may be called as secondary-treated wastewater.

Secondary treatment is the portion of a sewage treatment sequence removing dissolved and colloidal compounds measured as biochemical oxygen demand (BOD). Secondary treatment is traditionally applied to the liquid portion of sewage after primary treatment has removed settleable solids and floating material. Secondary treatment is typically performed by indigenous, aquatic microorganisms in a managed aerobic habitat. Bacteria and protozoa consume biodegradable soluble organic contaminants (e.g. sugars, fats, and organic short-chain carbon molecules from human waste, food waste, soaps and detergent) while reproducing to form cells of biological solids. Biological oxidation processes are sensitive to temperature and, between 0 °C and 40 °C, the rate of biological reactions increase with temperature. Most surface aerated vessels operate at between 4 °C and 32 °C.

Definitions

Primary Treatment

Primary treatment of sewage by quiescent settling allows separation of floating material and heavy solids from liquid waste. The remaining liquid usually contains less than half of the original solids content and approximately two-thirds of the BOD in the form of colloids and dissolved organic compounds. Where nearby water bodies can rapidly dilute this liquid waste, primary treated sewage may be discharged so natural biological decomposition oxidizes remaining waste.

The city of San Diego used Pacific Ocean dilution of primary treated effluent into the 21st century.

Secondary Treatment

The United States Environmental Protection Agency (EPA) defined secondary treatment based on the performance observed at late 20th-century bioreactors treating typical United States municipal sewage. Secondary treated sewage is expected to produce effluent with a monthly average of less than 30 mg/l BOD and less than 30 mg/l suspended solids. Weekly averages may be up to 50 percent higher. A sewage treatment plant providing both primary and secondary treatment is expected to remove at least 85 percent of the BOD and suspended solids from domestic sewage. The EPA regulations describe stabilization ponds as providing treatment equivalent to secondary treatment removing 65 percent of the BOD and suspended solids from incoming sewage and discharging approximately 50 percent higher effluent concentrations than modern bioreactors. The regulations also recognize the difficulty of meeting the specified removal percentages from combined sewers, dilute industrial wastewater, or Infiltration/Inflow.

Where natural waterways are too small to rapidly oxidize primary treated sewage, the liquid may be used to irrigate sewage farms until suburban property values encourage secondary treatment methods requiring less acreage. Glacial sand deposits allowed some northeastern United States cities to use intermittent sand filtration until more compact secondary treatment bioreactors became available.

Biological nutrient removal is regarded by some sanitary engineers as secondary treatment and by others as tertiary treatment. The differentiation may also differ from one country to another.

Tertiary Treatment

The purpose of tertiary treatment (also called "advanced treatment") is to provide a final treatment stage to further improve the effluent quality before it is discharged to the receiving environment (sea, river, lake, wet lands, ground, etc.). Tertiary treatment may include biological nutrient removal (alternatively, this can be classified as secondary treatment), disinfection and removal of micropollutants, such as environmental persistent pharmaceutical pollutants.

Process Upsets

Process upsets are temporary decreases in treatment plant performance caused by significant population change within the secondary treatment ecosystem. Conditions likely to create upsets include for example toxic chemicals and unusually high or low concentrations of organic waste BOD providing food for the bioreactor ecosystem.

Toxicity

Waste containing biocide concentrations exceeding the secondary treatment ecosystem tolerance level may kill a major fraction of one or more important ecosystem species. BOD reduction normally accomplished by that species temporarily ceases until other species reach a suitable population to utilize that food source, or the original population recovers as biocide concentrations decline.

Dilution

Waste containing unusually low BOD concentrations may fail to sustain the secondary treatment population required for normal waste concentrations. The reduced population surviving the starvation event may be unable to completely utilize available BOD when waste loads return to normal. Dilution may be caused by addition of large volumes of relatively uncontaminated water such as stormwater runoff into a combined sewer. Smaller sewage treatment plants may experience dilution from cooling water discharges, major plumbing leaks, firefighting, or draining large swimming pools.

A similar problem occurs as BOD concentrations drop when low flow increases waste residence time within the secondary treatment bioreactor. Secondary treatment ecosystems of college communities acclimated to waste loading fluctuations from student work/sleep cycles may have difficulty surviving school vacations. Secondary treatment systems accustomed to routine production cycles of industrial facilities may have difficulty surviving industrial plant shutdown. Populations of species feeding on incoming waste initially decline as concentration of those food sources decrease. Population decline continues as ecosystem predator populations compete for a declining population of lower trophic level organisms.

Peak Waste Load

High BOD concentrations initially exceed the ability of the secondary treatment ecosystem to utilize available food. Ecosystem populations of aerobic organisms increase until oxygen transfer limitations of the secondary treatment bioreactor are reached. Secondary treatment ecosystem populations may shift toward species with lower oxygen requirements, but failure of those species to use some food sources may produce higher effluent BOD concentrations. More extreme increases in BOD concentrations may drop oxygen concentrations before the secondary treatment ecosystem population can adjust, and cause an abrupt population decrease among important species. Normal BOD removal efficiency will not be restored until populations of aerobic species recover after oxygen concentrations rise to normal.

Design for Damage Control

Measures creating uniform wastewater loadings tend to reduce the probability of upsets. Fixed-film or attached growth secondary treatment bioreactors are similar to a plug flow reactor model circulating water over surfaces colonized by biofilm, while suspended-growth bioreactors resemble a continuous stirred-tank reactor keeping microorganisms suspended while water is being treated. Secondary treatment bioreactors may be followed by a physical phase separation to remove biological solids from the treated water. Upset duration of fixed film secondary treatment systems may be longer because of the time required to recolonize the treatment surfaces. Suspended growth ecosystems may be restored from a population reservoir. Activated sludge recycle systems provide an integrated reservoir if upset conditions are detected in time for corrective action. Sludge recycle may be temporarily turned off to prevent sludge washout during peak storm flows when dilution keeps BOD concentrations low. Suspended growth activated sludge systems can be operated in a smaller space than fixed-film trickling filter systems that treat the same amount of water; but fixed-film systems are better able to cope with drastic changes in the amount of biological material and can provide higher removal rates for organic material and suspended solids than suspended growth systems.

Wastewater flow variations may be reduced by limiting stormwater collection by the sewer system, and by requiring industrial facilities to discharge batch process wastes to the sewer over a time interval rather than immediately after creation. Discharge of appropriate organic industrial wastes may be timed to sustain the secondary treatment ecosystem through periods of low residential waste flow. Sewage treatment systems experiencing holiday waste load fluctuations may provide alternative food to sustain secondary treatment ecosystems through periods of reduced use. Small facilities may prepare a solution of soluble sugars. Others may find compatible agricultural wastes, or offer disposal incentives to septic tank pumpers during low use periods.

Process Types

Filter Beds (Oxidizing Beds)

In older plants and those receiving variable loadings, trickling filter beds are used where the settled sewage liquor is spread onto the surface of a bed made up of coke (carbonized coal), limestone chips or specially fabricated plastic media. Such media must have large surface areas to support the biofilms that form. The liquor is typically distributed through perforated spray arms. The distributed liquor trickles through the bed and is collected in drains at the base. These drains also provide a source of air which percolates up through the bed, keeping it aerobic. Biofilms of bacteria, protozoa and fungi form on the media's surfaces and eat or otherwise reduce the organic content. The filter removes a small percentage of the suspended organic matter, while the majority of the organic matter supports microorganism reproduction and cell growth from the biological oxidation and nitrification taking place in the filter. With this aerobic oxidation and nitrification, the organic solids are converted into biofilm grazed by insect larvae, snails, and worms which help maintain an optimal thickness. Overloading of beds may increase biofilm thickness leading to anaerobic conditions and possible bioclogging of the filter media and ponding on the surface.

Rotating Biological Contactors

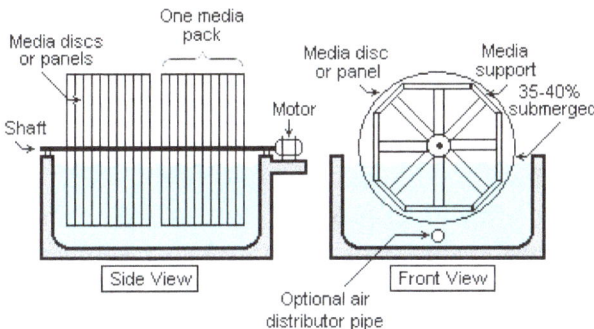

Schematic of a typical rotating biological contactor (RBC). The treated effluent clarifier/settler is not included in the diagram

Rotating biological contactors (RBCs) are robust mechanical fixed-film secondary treatment systems capable of withstanding surges in organic load. RBCs were first installed in Germany in 1960 and have since been developed and refined into a reliable operating unit. The rotating disks support the growth of bacteria and micro-organisms present in the sewage, which break down and stabilize organic pollutants. To be successful, micro-organisms need both oxygen to live and food to grow. Oxygen is obtained from the atmosphere as the disks rotate. As the micro-organisms grow, they build up on the media until they are sloughed off due to shear forces provided by the rotating discs in the sewage. Effluent from the RBC is then passed through a secondary clarifier where the sloughed biological solids in suspension settle as a sludge.

Activated Sludge

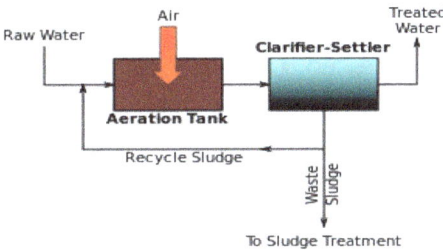

A generalized schematic of an activated sludge process.

Activated sludge is a common suspended-growth method of secondary treatment. Activated sludge plants encompass a variety of mechanisms and processes using dissolved oxygen to promote growth of biological floc that substantially removes organic material. Biological floc is an ecosystem of living biota subsisting on nutrients from the inflowing primary clarifier effluent. These mostly carbonaceous dissolved solids undergo aeration to be broken down and either biologically oxidized to carbon dioxide or converted to additional biological floc of reproducing micro-organisms. Nitrogenous dissolved solids (amino acids, ammonia, etc.) are similarly converted to biological floc or oxidized by the floc to nitrites, nitrates, and, in some processes, to nitrogen gas through denitrification. While denitrification is encouraged in some treatment processes, denitrification often impairs the settling of the floc causing poor quality effluent in many suspended aeration plants. Overflow from the activated sludge mixing chamber is sent to a secondary clarifier where the suspended biological floc settles out while the treated water moves into tertiary treatment or disinfection. Settled floc is returned to the mixing basin to continue growing in primary effluent. Like most ecosystems, population changes among activated sludge biota can reduce treatment efficiency. Nocardia, a floating brown foam sometimes misidentified as *sewage fungus*, is the best known of many different fungi and protists that can overpopulate the floc and cause process upsets. Elevated concentrations of toxic wastes including pesticides, industrial metal plating waste, or extreme pH, can kill the biota of an activated sludge reactor ecosystem.

Package Plants and Sequencing Batch Reactors

One type of system that combines secondary treatment and settlement is the cyclic activated sludge (CASSBR), or sequencing batch reactor (SBR). Typically, activated sludge is mixed with raw incoming sewage, and then mixed and aerated. The settled sludge is run off and re-aerated before a proportion is returned to the headworks.

The disadvantage of the CASSBR process is that it requires a precise control of timing, mixing and aeration. This precision is typically achieved with computer controls linked to sensors. Such a complex, fragile system is unsuited to places where controls may be unreliable, poorly maintained, or where the power supply may be intermittent. Extended aeration package plants use separate basins for aeration and settling, and are somewhat larger than SBR plants with reduced timing sensitivity.

Package plants may be referred to as *high charged* or *low charged*. This refers to the way the biological load is processed. In high charged systems, the biological stage is presented with a high organic load and the combined floc and organic material is then oxygenated for a few hours before being charged again with a new load. In the low charged system the biological stage contains a low organic load and is combined with flocculate for longer times.

Membrane Bioreactors

Membrane bioreactors (MBR) are activated sludge systems using a membrane liquid-solid phase separation process. The membrane component uses low pressure microfiltration or ultrafiltration membranes and eliminates the need for a secondary clarifier or filtration. The membranes are typically immersed in the aeration tank; however, some applications utilize a separate membrane tank. One of the key benefits of an MBR system is that it effectively overcomes the limitations associated with poor settling of sludge in conventional activated sludge (CAS) processes. The technology permits bioreactor operation with considerably higher mixed liquor suspended solids (MLSS) concentration than CAS systems, which are limited by sludge settling. The process is typically operated at MLSS in the range of 8,000–12,000 mg/L, while CAS are operated in the range of 2,000–3,000 mg/L. The elevated biomass concentration in the MBR process allows for very effective removal of both soluble and particulate biodegradable materials at higher loading rates. Thus increased sludge retention times, usually exceeding 15 days, ensure complete nitrification even in extremely cold weather.

The cost of building and operating an MBR is often higher than conventional methods of sewage treatment. Membrane filters can be blinded with grease or abraded by suspended grit and lack a clarifier's flexibility to pass peak flows. The technology has become increasingly popular for reliably pretreated waste streams and has gained wider acceptance where infiltration and inflow have been controlled, however, and the life-cycle costs have been steadily decreasing. The small footprint of MBR systems, and the high quality effluent produced, make them particularly useful for water reuse applications.

Aerobic Granulation

Aerobic granular sludge can be formed by applying specific process conditions that favour slow growing organisms such as PAOs (polyphosphate accumulating organisms) and GAOs (glycogen accumulating organisms). Another key part of granulation is selective wasting whereby slow settling floc-like sludge is discharged as waste sludge and faster settling biomass is retained. This process has been commercialized as Nereda process.

Surface-aerated Lagoons or Ponds

Aerated lagoons are a low technology suspended-growth method of secondary treatment using motor-driven aerators floating on the water surface to increase atmospheric oxygen transfer to the lagoon and to mix the lagoon contents. The floating surface aerators are typically rated to deliver the amount of air equivalent to 1.8 to 2.7 kg O_2/

kW·h. Aerated lagoons provide less effective mixing than conventional activated sludge systems and do not achieve the same performance level. The basins may range in depth from 1.5 to 5.0 metres. Surface-aerated basins achieve 80 to 90 percent removal of BOD with retention times of 1 to 10 days. Many small municipal sewage systems in the United States (1 million gal./day or less) use aerated lagoons.

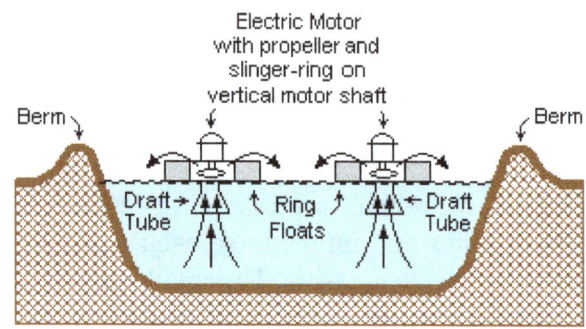

A TYPICAL SURFACE – AERATED BASIN

Note: The ring floats are tethered to posts on the berms.

A typical surface-aerated basin (using motor-driven floating aerators)

Constructed Wetlands

Primary clarifier effluent was discharged directly to eutrophic natural wetlands for decades before environmental regulations discouraged the practice. Where adequate land is available, stabilization ponds with constructed wetland ecosystems can be built to perform secondary treatment separated from the natural wetlands receiving secondary treated sewage. Constructed wetlands resemble fixed-film systems more than suspended growth systems, because natural mixing is minimal. Constructed wetland design uses plug flow assumptions to compute the residence time required for treatment. Patterns of vegetation growth and solids deposition in wetland ecosystems, however, can create preferential flow pathways which may reduce average residence time. Measurement of wetland treatment efficiency is complicated because most traditional water quality measurements cannot differentiate between sewage pollutants and biological productivity of the wetland. Demonstration of treatment efficiency may require more expensive analyses.

Emerging Technologies

- Biological Aerated (or Anoxic) Filter (BAF) or Biofilters combine filtration with biological carbon reduction, nitrification or denitrification. BAF usually includes a reactor filled with a filter media. The media is either in suspension or supported by a gravel layer at the foot of the filter. The dual purpose of this media is to support highly active biomass that is attached to it and to filter suspended solids. Carbon reduction and ammonia conversion occurs in aerobic mode and sometime achieved in a single reactor while nitrate conversion occurs in anoxic mode. BAF is operated either in upflow or downflow configuration depending on design specified by manufacturer.

- Integrated Fixed-Film Activated Sludge

- Moving Bed Biofilm Reactors typically requires smaller footprint than suspended-growth systems.

Electrochemical Treatment (ECT)

ECT process can be another alternative process for treating various wastewaters.

The major methods for ECT are: electro-coagulation (EC), electro-flotation (EF) and electro-oxidation (EO).

An ECT unit consists of anodes and cathodes in parallel mode. When electric power is applied from a power source, the anode material gets oxidized and the cathode is subjected to reduction of elemental metals and due to further reactions depending on conditions applied, removal of various pollutants takes place by EC and/or EF and/or EO mechanisms.

Electro-Flotation (EF)

EF is a simple process in which buoyant gases bubbles generated during electrolysis take along with them the pollutant materials to the surface of liquid body. The bubbles of hydrogen and oxygen which are generated from water electrolysis move upwards in the liquid phase. A layer of foam, containing gas bubbles and floated particles is formed at the surface of water. The rate of flotation depends on several parameters such as surface tension between the water particles and gas bubbles; the bubble size distribution and bubble density; size distribution of the particles; the residence time of the solution/liquid in the EC cell and the flotation tank; the particle and gas bubble zeta potentials; and the temperature, pH of the solution.

Electro-Oxidation (EO)

Decomposition of organic materials through EO treatment means the oxidation of organics present in wastewater to carbon dioxide and water or other oxides. The electrochemical oxidation of wastewater is achieved in two ways. First, by direct anodic oxidation, in which organics are adsorbed at the electrode and oxidized at the surface of the electrode and second, by indirect oxidation in which some oxidizing agents are generated electrochemically which are responsible for oxidation of organics present in the solution.

Organic pollutants are adsorbed on the anode surface in direct anodic oxidation process, where active oxygen (adsorbed hydroxyl radicals) or chemisorbed "active oxygen" is accountable for the oxidation of adsorbed Organics pollutants. The mechanism of oxidation of organic matter on oxide anode (MO_X) was suggested by Comninellis. The reactions involve are as follows:

$$H_2O + MO_X \rightarrow MO_X[OH] + H^+ + e^-$$

The adsorbed hydroxyl radicals may form chemisorbed active oxygen

$$MO_X[OH] \rightarrow MO_{X+1} + H^+ + e^-$$

The liberated chemisorbed active oxygen is responsible for the oxidation.

During the EO treatment process, two types, of oxidation is possible. In one way, toxic and non-biocompatible pollutants are converted into bio-degradable organics, so that further biological treatment can be initiated. In contrast, in other way, pollutants are oxidized to water and CO_2 and no further purification is necessary.

In an indirect oxidation process, strong oxidant such as hypochlorite/chlorine, ozone, and hydrogen peroxide are regenerated during electrolysis. Following reaction shows the formation of hypochlorite:

$$H_2O + Cl^- \rightarrow HOCl + H^+ + 2e^-$$

High voltage can led to formation of hydrogen peroxide and other molecules as follows:

$$H_2O \rightarrow OH, O, H^+, H_2O_2$$

These oxidants oxidize many inorganic and organic pollutants in the bulk solution.

Electro-coagulation (EC)

EC, like coagulation, is the process of destabilization of colloidal particles present in wastewater and can be achieved by two mechanisms: one in which an increase in ionic concentration, reduce the zeta potential, and adsorption of counter-ions on colloidal particles neutralises the particle charge; and other by well known mechanism of sweep flocculation.

Various reactions take place in the EC reactor during its operation. As the current is applied, the anode material undergoes oxidation and cathode gets reduced. If iron or Al electrodes are used, Fe^{2+} and Al^{3+} ion generation takes place at anode by the following reaction.

$$Fe(S) \rightarrow Fe^{2+}(aq) \quad + \quad 2e^-$$
$$Al(S) \rightarrow Al^{3+}(aq) \quad + \quad 3e^-$$

In addition, oxygen evolution can compete with iron or aluminum dissolution at the anode via the following reaction:

$$2H_2O(l) \rightarrow O_2(g) \quad + \quad 4H^+(aq) \quad + \quad 4e^-$$

At the cathode, hydrogen evolution takes place via the following reaction:

$$3H_2O(l) \quad + \quad 3e^- \rightarrow 3/2H_2(g) \quad + \quad 3OH^-(aq)$$

Liberated Fe^{2+}/Al^{3+} and OH^- ions react to form various monomeric and polymeric hydrolyzed species. The concentration of the hydrolyzed metal species depends on the metal concentration, and the solution pH. These metal hydrolysed products are responsible for the coagulation of pollutants from solution.

Factors Affecting Ect Process

Current density (J), electrolysis time (t) and anodic dissolution: Faraday's law describes the relationship between current density (J) and the amount of anode material that dissolves in the solution. It is given as:

$$w = \frac{MJt}{ZF}$$

Where, w is the theoretical amount of ion produced per unit surface area by current density J passed for duration of time, t. Z is the number of electrons involved in the oxidation/reduction reaction, M is the atomic weight of anode material and F is the Faraday's constant (96486 C/mol).

The pollutants removal efficiency depends directly on the concentration of aluminum ions produced by the metal electrodes, which in turn as per Faradays law depends upon the t and J. When the value of t and J increases, an increase occurs in the concentration of metal ions and their hydroxide flocs. Consequently, an increase in t and J increases the removal efficiency.

Theoretically, according to the Faraday's law when 1 F of charge passes through the circuit, 28 g of iron is dissolved at each electrode individually connected to the positive node of the power supply unit. During the coagulation process with iron electrodes, the valency of the coagulant increases, with Fe^{3+} being much more effective than the Fe^{2+}.

pH: The initial pH (pH_i) of the wastewater will have a significant impact on the efficiency of the ECT. The effects of pH_i on the ECT of wastewater can be reflected by the solubility of metal hydroxides. The effluent pH after ECT would increase. The incremental increase in pH with an incremental increase in the amount of current applied tends to decrease at higher current [10,11]. The general cause of the pH increase can be explained from the following equation:

$$2H_2O + 2e^- \rightarrow H_{2(g)} + 2OH^-(cathode)$$

At the cathode, generated hydrogen gas (which attaches to the flocculated agglomerates, resulting in flocs floation to the surface of the water) and this causes the pH to increase as the hydroxide-ion concentration in the water increases. This reaction is one of the dominant reactions that occur in the electro-flocculation system.

Also, due to the following reaction, pH is affected:

$$2H_2O \rightarrow O_2(g) + 4H^+ + 4e^-$$

These two reactions tend to neutralize pH. This is the reason, which, however, prevents larger pH increases due to larger hydroxides formations at higher current densities.

Conductivity and the effect of salts: Feed conductivity is an important parameter in ECT, since it directly affects the energy consumed per unit mass of pollutants removed. If conductivity is low, higher amount of energy is consumed per unit of mass of pollutants removed and vice versa. Due to this, some salts (commonly NaCl) are added to increase the conductivity of feed. When, salt is added to the solution, it reduces the solution resistance and hence, voltage distribution between the electrodes reduces. However, a too high conductivity may lead to secondary parasite reactions, diminishing the main reaction of the electrolytic decomposition. Additionally, the presence of chlorides can enhance the degradation of organic pollutants in the wastewaters due to the formation of various species (Cl_2, HOCl and ClO^-) formed as function of the pH. ClO^-, which is dominating at higher pH, has been reported as better oxidant among all chlorine species. Moreover, the type and concentration of salt also influences the effectiveness of the treatment. Salts of bi- and tri-valent metals are more effective than monovalent salts because of their high ionic strengths. Cl_2 and OH^- ions are generated on the surface of the anode and the cathode, respectively, when NaCl is used as an electrolyte in ECT. The organics are destroyed in the bulk solution by oxidation reaction of the regenerated oxidant. In an ECT cell, Cl_2/hypochlorite formation may take place because chloride is widely presented in many wastewaters.

References

- Johnson, D.L., S.H. Ambrose, T.J. Bassett, M.L. Bowen, D.E. Crummey, J.S. Isaacson, D.N. Johnson, P. Lamb, M. Saul, and A.E. Winter-Nelson (1997). "Meanings of environmental terms." Journal of Environmental Quality. 26: 581–589. doi:10.2134/jeq1997.00472425002600030002x

- Linsley, Ray K. & Franzini, Joseph B. Water-Resources Engineering (1972) McGraw-Hill ISBN 0-07-037959-9 pp.454–456

- Natural Disasters and Severe Weather. "Water Quality After a Tsunami". Centers for Disease Control and Prevention. Retrieved 2017-04-27

- Hanaor, Dorian A. H.; Sorrell, Charles C. (2014). "Sand Supported Mixed-Phase TiO_2 Photocatalysts for Water Decontamination Applications". Advanced Engineering Materials. 16 (2): 248–254. doi:10.1002/adem.201300259

- Dickens CWS and Graham PM. 2002. The Southern Africa Scoring System (SASS) version 5 rapid bioassessment for rivers "African Journal of Aquatic Science", 27:1–10

- G. Allen Burton, Jr., Robert Pitt (2001). Stormwater Effects Handbook: A Toolbox for Watershed Managers, Scientists, and Engineers. New York: CRC/Lewis Publishers. ISBN 0-87371-924-7

- "Center for Coastal Monitoring and Assessment: Mussel Watch Contaminant Monitoring". Ccma. nos.noaa.gov. 2014-01-14. Retrieved 2015-09-04

- International Water Management Institute, Colombo, Sri Lanka (2010). "Helping restore the quality of drinking water after the tsunami." Success Stories. Issue 7. doi:10.5337/2011.0030

- Di Luzio, Frank C. (January 1967). "Water Pollution Control: An American Must". Journal (Water Pollution Control Federation). Water Environment Federation. 39 (1): 1–7. JSTOR 25035710

- Goel, P.K. (2006). Water Pollution - Causes, Effects and Control. New Delhi: New Age International. p. 179. ISBN 978-81-224-1839-2

- Moss, Brian (2008). "Water Pollution by Agriculture" (PDF). Phil. Trans. Royal Society B. 363: 659–666. doi:10.1098/rstb.2007.2176

- International Organization for Standardization (ISO). "13.060: Water quality". Geneva, Switzerland. Retrieved 2011-07-04

- Kennish, Michael J. (1992). Ecology of Estuaries: Anthropogenic Effects. Marine Science Series. Boca Raton, FL: CRC Press. pp. 415–17. ISBN 978-0-8493-8041-9

- Lindsten, Don C. (September 1984). "Technology transfer: Water purification, U.S. Army to the civilian community". The Journal of Technology Transfer. 9 (1): 57–59. doi:10.1007/BF02189057

- Wachman, Richard (2007-12-09). "Water becomes the new oil as world runs dry". The Guardian. London. Retrieved 2015-09-23

- Zaikab, Gwyneth Dickey (2011-03-28). "Marine microbes digest plastic". Nature. Macmillan. ISSN 0028-0836. doi:10.1038/news.2011.191

- Haughey, A. (1968). "The Planktonic Algae of Auckland Sewage Treatment Ponds". New Zealand Journal of Marine and Freshwater Research. 2 (4): 721–766. doi:10.1080/00288330.1968.9515271

- Tchobanoglous, George; Burton, Franklin L.; Stensel, H. David; Metcalf & Eddy, Inc. (2003). Wastewater Engineering: Treatment and Reuse (4th ed.). McGraw-Hill. ISBN 0-07-112250-8

3

Biological Methods for Controlling Wastewater Pollution

Biological methods of controlling wastewater production can be divided into aerobic processes and anaerobic processes. Aerobic processes include aerated lagoons, stabilization ponds, aerobic digestion, etc. Anaerobic processes include anaerobic contact process, anaerobic digestion, etc. The major components used for controlling water pollution are discussed in this chapter.

Mechanical Biological Treatment

A mechanical biological treatment (MBT) system is a type of waste processing facility that combines a sorting facility with a form of biological treatment such as composting or anaerobic digestion. MBT plants are designed to process mixed household waste as well as commercial and industrial wastes.

Process

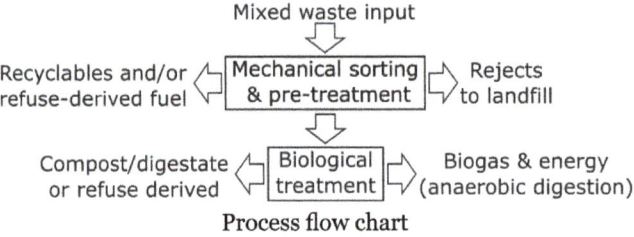

Process flow chart

The terms *mechanical biological treatment* or *mechanical biological pre-treatment* relate to a group of solid waste treatment systems. These systems enable the recovery of materials contained within the mixed waste and facilitate the stabilisation of the biodegradable component of the material.

The sorting component of the plants typically resemble a materials recovery facility. This component is either configured to recover the individual elements of the waste or produce a Refuse-derived fuel that can be used for the generation of power.

The components of the mixed waste stream that can be recovered include:

*   Ferrous Metal

- Non-ferrous metal

- Plastic

- Glass

Terminology

MBT is also sometimes termed BMT – biological mechanical treatment – however this simply refers to the order of processing, i.e. the biological phase of the system precedes the mechanical sorting. MBT should not be confused with MHT – *mechanical heat treatment*.

Mechanical Sorting

Wet material recovery facility, Hiriya, Israel

The "mechanical" element is usually an automated mechanical sorting stage. This either removes recyclable elements from a mixed waste stream (such as metals, plastics, glass and paper) or processes them. It typically involves factory style conveyors, industrial magnets, eddy current separators, trommels, shredders and other tailor made systems, or the sorting is done manually at hand picking stations. The mechanical element has a number of similarities to a materials recovery facility (MRF).

Some systems integrate a wet MRF to separate by density and flotation and to recover & wash the recyclable elements of the waste in a form that can be sent for recycling. MBT can alternatively process the waste to produce a high calorific fuel termed refuse derived fuel (RDF). RDF can be used in cement kilns or thermal combustion power plants and is generally made up from plastics and biodegradable organic waste. Systems which are configured to produce RDF include the Herhof and Ecodeco Processes. It is a common misconception that all MBT processes produce RDF. This is not the case and depends strictly on system configuration and suitable local markets for MBT outputs.

Biological Processing

Twin stage & UASB anaerobic digesters

The "biological" element refers to either:

- Anaerobic digestion

- Composting

- Biodrying

Anaerobic digestion harnesses anaerobic microorganisms to break down the biodegradable component of the waste to produce biogas and soil improver. The biogas can be used to generate electricity and heat.

Biological can also refer to a composting stage. Here the organic component is broken down by naturally occurring aerobic microorganisms. They breakdown the waste into carbon dioxide and compost. There is no green energy produced by systems employing only composting treatment for the biodegradable waste.

In the case of biodrying, the waste material undergoes a period of rapid heating through the action of aerobic microbes. During this partial composting stage the heat generated by the microbes result in rapid drying of the waste. These systems are often configured to produce a refuse-derived fuel where a dry, light material is advantageous for later transport and combustion.

Some systems incorporate both anaerobic digestion and composting. This may either take the form of a full anaerobic digestion phase, followed by the maturation (composting) of the digestate. Alternatively a partial anaerobic digestion phase can be induced on water that is percolated through the raw waste, dissolving the readily available sugars, with the remaining material being sent to a windrow composting facility.

By processing the biodegradable waste either by anaerobic digestion or by composting MBT technologies help to reduce the contribution of greenhouse gases to global warming.

Usable wastes for this system:

- Municipal solid waste

- Commercial and industrial waste

- Sewage sludge

Possible products of this system:

- Renewable fuel (biogas) leading to renewable power

- Recovered recycable materials such as metals, paper, plastics, glass etc.

- Digestate - an organic fertiliser and soil improver

- Carbon credits – additional revenues

- High calorific fraction refuse derived fuel - Renewable fuel content dependent upon biological component

- Residual unusable materials prepared for their final safe treatment (e.g. incineration or gasification) and/or landfill

Further advantages:

- Small fraction of inert residual waste

- Reduction of the waste volume to be deposited to at least a half (density > 1.3 t/m³), thus the lifetime of the landfill is at least twice as long as usually

- Utilisation of the leachate in the process

- Landfill gas not problematic as biological component of waste has been stabilised

- Daily covering of landfill not necessary

Consideration of Applications

MBT systems can form an integral part of a region's waste treatment infrastructure. These systems are typically integrated with kerbside collection schemes. In the event that a refuse-derived fuel is produced as a by-product then a combustion facility would be required. This could either be an incineration facility or a gasifier.

Alternatively MBT solutions can diminish the need for home separation and kerbside collection of recyclable elements of waste. This gives the ability of local authorities, municipalities and councils to reduce the use of waste vehicles on the roads and keep recycling rates high.

Position of Environmental Groups

Friends of the Earth suggests that the best environmental route for residual waste is to firstly maximise removal of remaining recyclable materials from the waste stream (such as metals, plastics and paper). The amount of waste remaining should be composted or anaerobically digested and disposed of to landfill, unless sufficiently clean to be used as compost.

A report by Eunomia undertook a detailed analysis of the climate impacts of different residual waste technologies. It found that an MBT process that extracts both the metals and plastics prior to landfilling is one of the best options for dealing with our residual waste, and has a lower impact than either MBT processes producing RDF for incineration or incineration of waste without MBT.

Friends of the Earth does not support MBT plants that produce refuse derived fuel (RDF), and believes MBT processes should occur in small, localised treatment plants.

Types of Biological Processes for Wastewater Treatment

For the treatment of wastewater the principle biological processes are divided into two categories: suspended growth and attached growth processes.

[A] Suspended Growth Processes

In this process the microorganisms responsible for treatment are maintained in liquid suspension by mixing methods. The following section describes the two most widely used suspended growth processes activated sludge and aerated lagoons and one recently introduced membrane bioreactors.

Activated Sludge Process: It is the most widely used process for wastewater treatment. It consists of two sets of basins. In the first, air is pumped through perforated pipes at the bottom of the basin, air rises through the water in the form of many small bubbles. These bubbles provide oxygen from the air to the water and create highly turbulent conditions that favor intimate contact between cells, the organic material in the water and oxygen. The second basin is a settling tank where water flow is made to be very quiet so that the cellular material is removed by gravitational settling. Some of the cell material collected at the bottom is captured and fed back into the first basin to seed the process. The rest of the sludge is taken for anaerobic digestion.

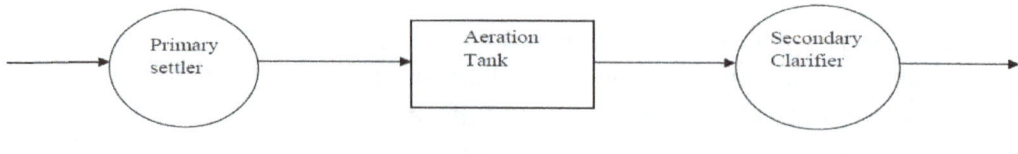

Activated Sludge Process

Aerated Lagoons and Oxidative Pond: Oxidative ponds are shallow ponds with a depth of 1 to 2 m where primary treated waste is decomposed by the microorganisms. Oxidative ponds maintain aerobic conditions, the decomposition near the surface is aerobic whereas near the bottom is anaerobic. They have a mix of conditions and are called facultative ponds. The oxygen required for decomposition is derived from either surface aeration or the photosynthesis of algae.

Membrane Bioreactors: These membranes have been designed to reduce the size of secondary treatment tanks and improve the separation efficiency. They draw water from mixed liquor into hollow fiber membranes which have a pore size of about 0.2µm. The membranes are submerged in the activated sludge aeration tank and there is no need of a secondary clarifier or they may be present outside the aeration zone.

[B] Attached Growth Treatment

In this treatment, the microorganisms that are used for the conversion of nutrients or organic material are attached to the inert packing material. The organic material is removed from the wastewater flowing past the biofilm or the attached growth. Sand, gravel, rock and a wide variety of plastic and other synthetic material is used as the packing material. They can be used both as aerobic when partially submerged in wastewater or as anaerobic when fully submerged and no air space above it.

Trickling Filter: This is the most widely used attached growth process. It consists of a rotating distribution arm that sprays wastewater above the bed of plastic material or other coarse material. The spacing between the packing allows air to easily circulate so that aerobic conditions are present. The media in the bed is covered by a layer of biological slime containing bacteria, fungi etc that adsorbs and consumes the waste trickling through the bed.

Rotating Biological Contactors: It consists of a series of closely spaced circular plastic disks that are attached to a rotating hydraulic shaft. 40% of the bottom of each plate is dipped in the wastewater and the film which grows on the disk moves in and out of the wastewater.

Aerobic Treatment System

An aerobic treatment system or ATS, often called (incorrectly) an aerobic septic system, is a small scale sewage treatment system similar to a septic tank system, but which uses an aerobic process for digestion rather than just the anaerobic process used in septic systems. These systems are commonly found in rural areas where public sewers are not available, and may be used for a single residence or for a small group of homes.

Unlike the traditional septic system, the aerobic treatment system produces a high quality secondary effluent, which can be sterilized and used for surface irrigation. This allows much greater flexibility in the placement of the leach field, as well as cutting the required size of the leach field by as much as half.

Process

The ATS process generally consists of the following phases:

- Pre-treatment stage to remove large solids and other undesirable substances.

- Aeration stage, where aerobic bacteria digest biological wastes.

- Settling stage allows undigested solids to settle. This forms a sludge that must be periodically removed from the system.

- Disinfecting stage, where chlorine or similar disinfectant is mixed with the water, to produce an antiseptic output.

The disinfecting stage is optional, and is used where a sterile effluent is required, such as cases where the effluent is distributed above ground. The disinfectant typically used is tablets of calcium hypochlorite, which are specially made for waste treatment systems. The tablets are intended to break down quickly in sunlight. Stabilized forms of chlorine persist after the effluent is dispersed, and can kill plants in the leach field.

Since the ATS contains a living ecosystem of microbes to digest the waste products in the water, excessive amounts of items such as bleach or antibiotics can damage the ATS environment and reduce treatment effectiveness. Non-digestible items should also be avoided, as they will build up in the system and require more frequent sludge removal.

Types of Aerobic Treatment Systems

Small scale aerobic systems generally use one of two designs, fixed-film systems, or continuous flow, suspended growth aerobic systems (CFSGAS). The pre-treatment and effluent handling are similar for both types of systems, and the difference lies in the aeration stage.

Fixed Film Systems

Fixed film systems use a porous medium which provides a bed to support the biomass film that digests the waste material in the wastewater. Designs for fixed film systems vary widely, but fall into two basic categories (though some systems may combine both methods). The first is a system where the media is moved relative to the wastewater, alternately immersing the film and exposing it to air, while the second uses a stationary media, and varies the wastewater flow so the film is alternately submerged and exposed to air. In both cases, the biomass must be exposed to both wastewater and air for the aerobic digestion to occur. The film itself may be made of any suitable porous material, such as formed plastic or peat moss. Simple systems use stationary media, and rely on intermittent, gravity driven wastewater flow to provide periodic exposure to air and wastewater. A common moving media system is the rotating biological contactor (RBC), which uses disks rotating slowly on a horizontal shaft. Approximately 40 percent of the disks are submerged at any given time, and the shaft rotates at a rate of one or two revolutions per minute.

Continuous Flow, Suspended Growth Aerobic Systems

CFSGAS systems, as the name implies, are designed to handle continuous flow, and do not provide a bed for a bacterial film, relying rather on bacteria suspended in the wastewater. The suspension and aeration are typically provided by an air pump, which pumps air through the aeration chamber, providing a constant stirring of the wastewater in addition to the oxygenation. A medium to promote fixed film bacterial growth may be added to some systems designed to handle higher than normal levels of biomass in the wastewater.

Retrofit or Portable Aerobic Systems

Another increasingly common use of aerobic treatment is for the remediation of failing or failed anaerobic septic systems, by retrofitting an existing system with an aerobic feature. This class of product, known as aerobic remediation, is designed to remediate biologically failed and failing anaerobic distribution systems by significantly reducing the biochemical oxygen demand (BOD5) and total suspended solids (TSS) of the effluent. The reduction of the BOD5 and TSS reverses the developed bio-mat. Further, effluent with high dissolved oxygen and aerobic bacteria flow to the distribution component and digest the bio-mat.Doing so on single tank systems where solids do not have anywhere to settle, or there is no a clarifying area can do damage to the field lines as the solid matter is stirred up in the tank.

Composting Toilets

Composting toilets are designed to treat only toilet waste, rather than general residential waste water, and are typically used with water-free toilets rather than the flush toilets associated with the above types of aerobic treatment systems. These systems treat the waste as a moist solid, rather than in liquid suspension, and therefore separate urine from feces during treatment to maintain the correct moisture content in the system. An example of a composting toilet is the clivus multrum, which consists of an inclined chamber that separates urine and feces and a fan to provide positive ventilation and prevent odors from escaping through the toilet. Within the chamber, the urine and feces are independently broken down not only by aerobic bacteria, but also by fungi, arthropods, and earthworms. Treatment times are very long, with a minimum time between removals of solid waste of a year; during treatment the volume of the solid waste is decreased by 90 percent, with most being converted into water vapor and carbon dioxide. Pathogens are eliminated from the waste by the long durations in inhospitable conditions in the treatment chamber.

Comparison to Traditional Septic Systems

The aeration stage and the disinfecting stage are the primary differences from a traditional septic system; in fact, an aerobic treatment system can be used as a secondary treatment for septic tank effluent. These stages increase the initial cost of the aerobic system, and also the maintenance requirements over the passive septic system. Unlike many other biofilters, aerobic treatment systems require a constant supply of electricity to drive the air pump increasing overall system costs. The disinfectant tablets must be periodically replaced, as well as the electrical components (air compressor) and mechanical components (air diffusers). On the positive side, an aerobic system produces a higher quality effluent than a septic tank, and thus the leach field can be smaller than that of a conventional septic system, and the output can be discharged in areas too environmentally sensitive for septic system output. Some aerobic systems recycle the effluent through a sprinkler system, using it to water the lawn where regulations approve.

Effluent Quality

Since the effluent from an ATS is often discharged onto the surface of the leach field, the quality is very important. A typical ATS will, when operating correctly, produce an effluent with less than 30 mg/liter BOD5, 25 mg/L TSS, and 10,000 cfu/mL fecal coliform bacteria. This is clean enough that it cannot support a biomat or "slime" layer like a septic tank.

ATS effluent is relatively odorless; a properly operating system will produce effluent that smells musty, but not like sewage. Aerobic treatment is so effective at reducing odors, that it is the preferred method for reducing odor from manure produced by farms.

Basic Equation for Biological Treatment of Wastewater

According to Monod kinetics,

$$\mu = \mu_{max}\left[\frac{s}{k_s + s}\right]$$

Where, s is the substrate concentration, ks is the substrate concentration when $\mu(=\mu_{max}/2)$, μ_{max} is the maximum μ when substrate is not limiting.

Also, solid production rate $\left(\dfrac{dX}{dt}\right)$ is related to substrate utilization rate $\left(\dfrac{dS}{st}\right)$ following relationship:

$$\left(\frac{dX}{dt}\right) = Y\left(\frac{dS_r}{st}\right)$$

Where, S_r is the mass of soluble substrate (i.e. BOD), Y is the yield coefficient (kg of new cells formed/kg BOD removed). However owing to large treatment time in many of the large treatment units, substantial number of cells may die because of endogenous respiration. Therefore,

Net production rate $\left(\dfrac{dX}{dt}\right) = Y\left(\dfrac{dS_r}{dt}\right) - K_d X$

Where, K_d is the endogenous respiration decay rate constant.

For growth phase only,

$$\left(\frac{dS_r}{dt}\right) = Y\left(\frac{dX}{dt}\right) = \frac{\mu}{Y}X = \frac{\mu_{max}}{Y}X\left(\frac{s}{k_s + s}\right)$$

Case 1- $S >> K_s$

$$\frac{dS_r}{dt} \approx KX \quad \text{where} \quad K = \frac{\mu_{max}}{Y}$$

i.e. removal rate is independent of substrate concentration and that the removal rate depends on X only.

Case 2- $S << K_s$

$$\frac{dS_r}{dt} = K \times \frac{S}{K_s} \approx K' \times S, K' = \frac{K}{K_s} = \frac{\mu_{max}}{YK_s}$$

Here, removal rate depends both upon X and S. Where, X is the mass of biomass in the system (usually represented by MLSS i.e. Mixed Liquor Suspended Solid), μ is the specific growth rate constant (time^{-1}).

Major Terms

[a] Hydraulic detention time, $t_{HRT} = \dfrac{V}{Q} = \dfrac{\text{volume}}{\text{flowrate}}$

[b] Sludge age or mean residence time θ_c

$$\theta_C = \frac{\text{Mass of solid in the system}}{\text{Mass of solid leaving system per day}} = \frac{xV}{x'Q} = \frac{x}{x'}t_{HRT}$$

Where, x (=X/V) the concentration of microbial solution in the system, x' is the concentration of solids withdrawn.

For the flow through system, $x' = x$ and $\theta_C = t_{HRT}$

For the flow system with recycling, $x' < x$ and $\theta_C \rightarrow t_{HRT}$

[c] Food to microorganism ratio

$$\frac{F}{M} = \frac{\text{Substrate removal rate}}{\text{Solid(microorganisms) in the system}} = \frac{(S_o - S)/t}{X} = \frac{S_o - S}{xVt}$$

Activated Sludge

Activated sludge tank at Beckton sewage treatment plant, UK - the white bubbles are due to the diffused air aeration system

The activated sludge process is a process for treating sewage and industrial wastewaters using air and a biological floc composed of bacteria and protozoa.

Purpose

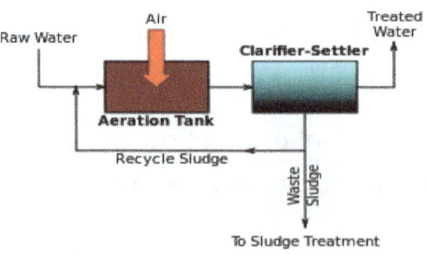

A generalized, schematic diagram of an activated sludge process

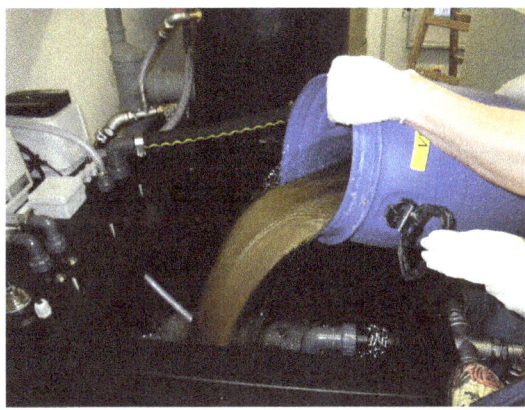

Activated sludge addition (seeding) to a pilot scale membrane bioreactor in Germany

In a sewage (or industrial wastewater) treatment plant, the activated sludge process is a biological process that can be used for one or several of the following purposes: oxidizing carbonaceous biological matter, oxidizing nitrogenous matter: mainly ammonium and nitrogen in biological matter, removing nutrients (nitrogen and phosphorus).

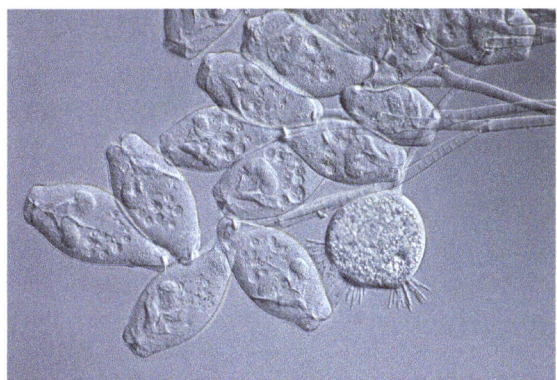

Activated sludge under the microscope

Process Description

The general arrangement of an activated sludge process for removing carbonaceous pollution includes the following items:

- Aeration tank where air (or oxygen) is injected in the mixed liquor.

- Settling tank (usually referred to as "final clarifier" or "secondary settling tank") to allow the biological flocs (the sludge blanket) to settle, thus separating the biological sludge from the clear treated water.

Treatment of nitrogenous matter or phosphorus involves additional steps where the mixed liquor is left in anoxic condition (meaning that there is no residual dissolved oxygen) and anaerobic zones (meaning that next to dissolved oxygen also Nitrite and Nitrate are absent).

Bioreactor and Final Clarifier

The process involves air or oxygen being introduced into a mixture of screened, and primary treated sewage or industrial wastewater (wastewater) combined with organisms to develop a biological floc which reduces the organic content of the sewage. This material, which in healthy sludge is a brown floc, is largely composed of saprotrophic bacteria but also has an important protozoan flora component mainly composed of amoebae, Spirotrichs, Peritrichs including Vorticellids and a range of other filter-feeding species. Other important constituents include motile and sedentary Rotifers. In poorly managed activated sludge, a range of mucilaginous filamentous bacteria can develop including *Sphaerotilus natans* which produces a sludge that is difficult to settle and can result in the sludge blanket decanting over the weirs in the settlement tank to severely contaminate the final effluent quality. This material is often described as sewage fungus but true fungal communities are relatively uncommon.

The combination of wastewater and biological mass is commonly known as *mixed liquor*. In all activated sludge plants, once the wastewater has received sufficient treatment, the mixed liquor is settled in settling tanks and the treated or purified supernatant is run off to discharge or sometimes undergoes further treatment before discharge or reuse. The majority of the settled material, the sludge, is returned to the head of the aeration system to re-seed the new wastewater entering the tank. This fraction of the floc is called *return activated sludge* (RAS).

The space required for a sewage treatment plant can be reduced by using a membrane bioreactor to remove some wastewater from the mixed liquor prior to treatment. This results in a more concentrated waste product that can then be treated using the activated sludge process.

Many sewage treatment plants use axial flow pumps to transfer nitrified mixed liquor from the aeration zone to the anoxic zone for denitrification. These pumps are often referred to as internal mixed liquor recycle pumps (IMLR pumps). The raw sewage, the RAS, and the nitrified mixed liquor are mixed by submersible mixers in the anoxic zones in order to achieve denitrification.

Sludge Production

Activated sludge is also the name given to the active biological material produced by activated sludge plants. Excess sludge is called "surplus activated sludge" or "waste activated sludge" or "excess aerobic biomass" and is removed from the treatment process to keep the ratio of biomass to food supplied in the wastewater in balance. This sewage sludge can be mixed with primary sludge from the primary clarifiers and undergoes further sludge treatment for example by anaerobic digestion, followed by thickening, dewatering, composting and land application. Today more often primary and secondary sludge are thickened separately before mixed towards digestion.

The amount of sewage sludge produced from the activated sludge process is directly proportional to the amount of wastewater treated, but more specific proportional to the load of pollutants in kg COD or BOD. The total sludge production consists of the sum of primary sludge from the primary sedimentation tanks as well as waste activated sludge from the bioreactors. A typical activated sludge process for domestic wastewater produces about 70–100 g/m^3 of waste activated sludge (that is g of dry solids produced per m3 of wastewater treated). A value of 80 g/m^3 is regarded as being typical. In addition, about 110–170 g/m^3 of primary sludge is produced in the primary sedimentation tanks which most - but not all - of the activated sludge process configurations use.

A variant of the activated sludge process is the Nereda process where aerobic granular sludge is developed by applying specific process conditions that favour slow growing organisms.

Process Control

The general method to do this is to monitor sludge blanket level, the biomass concentration (MLSS in g/l), the Sludge Volume after 30 minutes settling (SV30 in ml/l), SVI (Sludge Volume Index), MCRT (Mean Cell Residence Time) or SRT (Sludge Retention Time in days), F/M (Food to Microorganism), as well as the biota of the activated sludge and the major nutrients DO (Dissolved oxygen), nitrogen both in reduced (NH$_4$-N) and oxidezed forms (NO$_2$-N and NO$_3$-N), phosphorus, COD (Chemical oxygen demand) and BOD (Biochemical oxygen demand).

In the reactor/aerator + clarifier system:

- The sludge blanket is measured from the bottom of the clarifier to the level of settled solids in the clarifier's water column; this, in large plants, can be done up to three times a day.

- The SVI is the volume of settled sludge in milliliters occupied by 1 gram of dry sludge solids after 30 minutes of settling in a 1000 milliliter graduated cylinder.

- The MCRT is the total mass (lbs) of mixed liquor suspended solids in the aerator and clarifier divided by the mass flow rate (lbs/day) of mixed liquor suspended solids leaving as WAS and final effluent.

- The F/M is the ratio of food fed to the microorganisms each day to the mass of microorganisms held under aeration. Specifically, it is the amount of BOD fed to the aerator (lbs/day) divided by the amount (lbs) of Mixed Liquor Volatile Suspended Solids (MLVSS) under aeration. Note: Some references use MLSS (Mixed Liquor Suspended Solids) for expedience, but MLVSS is considered more accurate for the measure of microorganisms. Again, due to expedience, COD is generally used, in lieu of BOD, as COD is a rather easy and reliable analyzing method and as BOD takes five days for results.

Based on these control methods, the amount of settled solids in the mixed liquor can be varied by wasting activated sludge (WAS) or returning activated sludge (RAS).

Types of Plants

There are a variety of types of activated sludge plants. These include:

Package Plants

There are a wide range of types of package plants, often serving small communities or industrial plants that may use hybrid treatment processes often involving the use of aerobic sludge to treat the incoming (and preference settled) sewage. In such plants the primary settlement stage of treatment may be omitted (PM I do not agree (JR) removal of solids is important for package or attached biofilm processes). In these plants, a bi-otic floc is created which provides the required substrate. Package plants are designed and fabricated by specialty engineering firms in dimensions that allow for their transportation to the job site in public highways, typically width and height of 12 by 12 feet. Length varies with capacity with larger plants being fabricated in pieces and welded on site. Steel is preferred over synthetic materials (e.g., plastic) for its durability.

Package plants are commonly variants of extended aeration, to promote the 'fit & forget' approach required for small communities without dedicated operational staff. There are various standards to assist with their design.

Oxidation Ditch

In some areas, where more land is available, sewage is treated in large round or oval ditches with one or more horizontal aerators typically called brush or disc aerators which drive the mixed liquor around the ditch and provide aeration. These are oxidation ditches, often referred to by manufacturer's trade names such as Pasveer, Orbal or Carrousel. They have the advantage that they are relatively easy to maintain and are resilient to shock loads that often occur in smaller communities (i.e. at breakfast time and in the evening).

Oxidation ditches are installed commonly as 'fit & forget' technology, with typical design parameters of a hydraulic retention time of 24 – 48 hours, and a Sludge Retention Time or sludge age of 12 – 20 days. This compares with nitrifying activated sludge plants having a retention time of 8 hours, and a sludge age of 8 – 12 days.

Deep Shaft/Vertical Treatment

Where land is in short supply sewage may be treated by injection of oxygen into a pressured return sludge stream which is injected into the base of a deep columnar tank buried in the ground. Such shafts may be up to 100 metres deep and are filled with sewage liquor. As the sewage rises the oxygen forced into solution by the pressure at the base of the shaft breaks out as molecular oxygen providing a highly efficient source of oxygen for the activated sludge biota. The rising oxygen and injected return sludge provide the physical mechanism for mixing of the sewage and sludge. Mixed sludge and sewage is decanted at the surface and separated into supernatant and sludge components. The efficiency of deep shaft treatment can be high.

Surface aerators are commonly quoted as having an aeration efficiency of 0.5 - 1.5 kg O_2/kWh, diffused aeration as 1.5 - 2.5 kg O_2/kWh. Deep Shaft claims 5 – 8 kg O_2/kWh.

However, the costs of construction are high. Deep Shaft has seen the greatest uptake in Japan, because of the land area issues. Deep Shaft was developed by ICI, as a spin-off from their Pruteen process. In the UK it is found at three sites: Tilbury, Anglian water, treating a wastewater with a high industrial contribution; Southport, United Utilities, because of land space issues; and Billingham, ICI, again treating industrial effluent, and built (after the Tilbury shafts) by ICI to help the agent sell more.

DeepShaft is a patented, licensed, process. The licensee has changed several times and currently (2015) Noram Engineering sells it.

Surface-aerated Basins

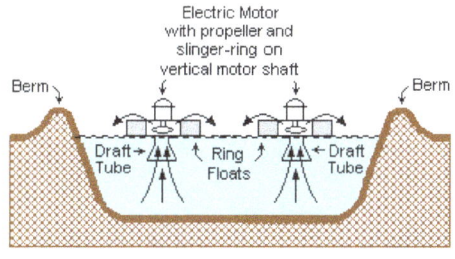

A TYPICAL SURFACE – AERATED BASIN

Note: The ring floats are tethered to posts on the berms.

A Typical Surface-Aerated Basing (using motor-driven floating aerators)

Most biological oxidation processes for treating industrial wastewaters have in common the use of oxygen (or air) and microbial action. Surface-aerated basins achieve 80 to 90% removal of BOD with retention times of 1 to 10 days. The basins may range in depth from 1.5 to 5.0 metres and utilize motor-driven aerators floating on the surface of the wastewater.

In an aerated basin system, the aerators provide two functions: they transfer air into the basins required by the biological oxidation reactions, and they provide the mixing required for dispersing the air and for contacting the reactants (that is, oxygen, wastewater and microbes). Typically, the floating surface aerators are rated to deliver the amount of air equivalent to 1.8 to 2.7 kg O_2/kWh. However, they do not provide as good mixing as is normally achieved in activated sludge systems and therefore aerated basins do not achieve the same performance level as activated sludge units.

Biological oxidation processes are sensitive to temperature and, between 0 °C and 40 °C, the rate of biological reactions increase with temperature. Most surface aerated vessels operate at between 4 °C and 32 °C.

Sequencing Batch Reactors (SBRs)

Sequencing batch reactors (SBRs) treat wastewater in batches within the same vessel. This means that the bioreactor (and the specific aerobic, anoxic and anaerobic circumstances) and final clarifier are not separated in space but in a timed sequence. The installation consists of at least two identically equipped tanks with a common inlet, which can be switched between them. While one tank is in settle/decant mode the other is aerating and filling.

Aeration Methods

Diffused Aeration

Sewage liquor is run into deep tanks with diffuser grid aeration systems that are attached to the floor. These are like the diffused airstone used in tropical fish tanks but on a much larger scale. Air is pumped through the blocks and the curtain of bubbles formed both oxygenates the liquor and also provides the necessary mixing action. Where capacity is limited or the sewage is unusually strong or difficult to treat, oxygen may be used instead of air. Typically, the air is generated by some type of blower or compressor.

Surface Aerators (Cones)

Vertically mounted tubes of up to 1-metre diameter extending from just above the base of a deep concrete tank to just below the surface of the sewage liquor. A typical shaft might be 10 metres high. At the surface end, the tube is formed into a cone with helical vanes attached to the inner surface. When the tube is rotated, the vanes spin liquor up and out of the cones drawing new sewage liquor from the base of the tank. In many works, each cone is located in a separate cell that can be isolated from the remaining cells if required for maintenance. Some works may have two cones to a cell and some large works may have 4 cones per cell.

Pure Oxygen Aeration

Pure oxygen activated sludge aeration systems are sealed-tank reactor vessels with surface aerator type impellers mounted within the tanks at the oxygen atmosphere-mixed liquor surface interface. The amount of oxygen entrainment, or DO (Dissolved Oxygen), can be controlled by a weir adjusted level control, and a vent gas oxygen controlled oxygen feed valve. Oxygen is generated on site by cryogenic distillation of air, pressure swing adsorption, or other methods. These systems are used where wastewater plant space is at a premium and high sewage throughput is required as high energy costs are involved in purifying oxygen.

Recent Developments

A new development of the activated sludge process is the Nereda process which produces a granular sludge that settles very well (the sludge volume index is reduced from 200-300 to 40 mL/g). A new process reactor system is created to take advantage of this quick settling sludge and is integrated into the aeration tank instead of having a separate unit outside. About 30 Nereda wastewater treatment plants worldwide are operational, under construction or under design, varying in size from 5,000 up to 858,000 person equivalent.

History

The Davyhulme Sewage Works Laboratory, where the activated sludge process was developed in the early 20th century

The activated sludge process was discovered in 1913 in the United Kingdom by two engineers, Edward Ardern and W.T. Lockett, who were conducting research for the Manchester Corporation Rivers Department at Davyhulme Sewage Works. This development led to arguably the single most significant improvement in public health and the environment during the course of the century.

In 1912, Dr. Gilbert Fowler, a scientist at the University of Manchester, observed experiments being conducted at the Lawrence Experiment Station at Massachusetts involving the aeration of sewage in a bottle that had been coated with algae. Fowler's engineering colleagues, Ardern and Lockett, experimented on treating sewage in a draw-and-fill reactor, which produced a highly treated effluent. They aerated the waste-water continuously for about a month and were able to achieve a complete nitrification of the sample material. Believing that the sludge had been activated (in a similar manner to activated carbon) the process was named *activated sludge*. Not until much later was it realized that what had actually occurred was a means to concentrate biological organisms, decoupling the liquid retention time (ideally, low, for a compact treatment system) from the solids retention time (ideally, fairly high, for an effluent low in BOD_5 and ammonia.)

Their results were published in their seminal 1914 paper, and the first full-scale continuous-flow system was installed at Worcester two years later. In the aftermath of the First World War the new treatment method spread rapidly, especially to the USA, Denmark, Germany and Canada. By the late 1930s, the activated sludge treatment became a well-known biological wastewater treatment process in those countries where sewer systems and sewage treatment plants were common.

Activated Sludge Process

Activated sludge process is used during secondary treatment of wastewater. Activated sludge is a mixture of bacteria, fungi, protozoa and rotifers maintained in suspension by aeration and mixing.

In this process, a biomass of aerobic organisms is grown in large aerated basins. These organisms breakdown the waste and use it as their food to grow themselves.

Activated sludge processes return settled sludge to the aeration basins in order to maintain the right amount of organisms to handle the incoming "food".

Activated sludge processes have removal efficiencies in the range (95-98%) than trickling filters (80-85%).

Working of Activated Sludge System

- A primary settler (or primary clarifier) may be introduced to remove part of the suspended solids present in the influent and this reduces the organic load to the activated sludge system.

- The biological reactor or aeration tank is filled with a mixture of activated sludge and influent, known as "mixed liquor". It is necessary to maintain certain mixed liquor suspended solid (MLSS) in the aerated tank maintain good removal efficiency.

- The aeration equipment transfers the oxygen necessary for the oxidation of organic material into the reactor, while simultaneously introducing enough turbulence to keep the sludge flocs in suspension.

- The continuous introduction of new influent results in a continuous discharge of mixed liquor to the secondary settler where separation of solids and liquid takes place.

- The liquid leaves the system as treated effluent, whereas some part of the sludge is recirculated to the aeration tank called as 'return sludge' and rest of sludge is taken for anaerobic digestion.

Designing of Activated Sludge System

Suppose, Q is the flow rate of influent (m³/d), Q_W is the flow rate of waste sludge (m³/d), Q_r is the flow rate of return activated sludge (m³/d), V is the volume of aeration tank (m³), S_o is the influent soluble substrate concentration (BOD g/m³), S is the effluent soluble substrate concentration (BOD g/m³), X_o is the concentration of biomass in influent (g VSS/m³), X_R is the concentration of biomass in return line from clarifier (g VSS/m³), X_r is the concentration of biomass in sludge drain (g VSS/m³) and X_e is the concentration of biomass in effluent (g VSS/m³). VSS stands for volatile suspended solids.

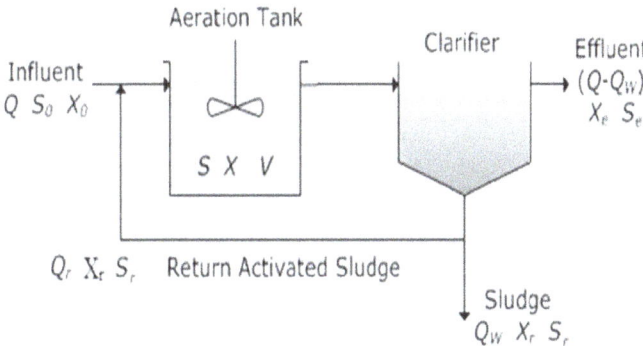

Activated sludge process

(a) Equations used for design of aeration tank

$$\theta_C = \frac{VX}{Q_W X_r}$$

$$\frac{1}{\theta_C} = \frac{QY(S_0 - S)}{VX} - k_d$$

(b) Mass balance around clarifier

$$X(Q + Q_r) = (Q - Q_W)X_e + (Q_W + Q_r)X_r$$

For $X_e = 0$

$$X(Q + Q_r) = (Q_W + Q_r)X_r$$

$$Q_r = \frac{QX - Q_W X_r}{X_r - X}$$

$$\text{Re cycle ratio} = \frac{Q_r}{Q}$$

Problem: An activated-sludge system is to be used for secondary treatment of 15,000 m³/d of municipal wastewater. After primary clarification, the BOD is 170 mg/L, and it is desired to have not more than 25 mg/L of soluble BOD in the effluent. A completely mixed reactor is to be used, and pilot-plant analysis has established the following values: hydraulic detention time (θ_c)=10 d yield coefficient (Y)=0.5 kg/kg, k_d=0.05 d⁻¹. Assuming an MLSS concentration of 4500 mg/L and an underflow concentration of 12,000 mg/L from the secondary clarifier, determine (1) the volume of the reactor, (2) the mass and volume of solids that must be wasted each day, and (3) the recycle ratio.

Solution: Given that Q=10,000 m³/d, θ_c =10 d

Using

$$\frac{1}{\theta_C} = \frac{QY(S_0 - S)}{VX} - k_d$$

$$0.1 d^{-1} = \frac{15,000 \, m^3/d \left(0.17 \, kg/m^3 - 0.025 \, kg/m^3\right)}{V \times 4.5 \, kg/m^3} - 0.05 d^{-1}$$

$$V = 1611 m^3$$

$$\text{Using } \theta_C = \frac{VX}{Q_W X_r} = 10$$

$$Q_W X_r = 724.95 \text{ kg/d}$$

If the concentration of solids in the underflow is 12,000 mg/L

$$Q_W = \frac{724.95 \text{ kg/d}}{12 \text{ kg/m}^3} = 60.41 m^3/d$$

For X_e=0

$$Q_r = \frac{QX - Q_W X_r}{X_r - X} = \frac{15,000 \, m^3/d \times 4.5 kg/m^3 - 724.95 kg/d}{12 kg/m^3 - 4.5 kg/m^3} = 8903.34 \, m^3/d$$

$$\text{Re cycle ratio} = \frac{Q_r}{Q} = \frac{8903.34}{15,000} = 0.59$$

Ponds and Lagoons

Other than activated sludge processes, ponds and lagoons are most common suspended- culture biological systems used for the treatment of wastewater.

A wastewater pond, alternatively known as a stabilization pond, oxidation pond, and sewage lagoon, consists of a large, shallow earthen basin in which wastewater is retained long enough for natural purification processes.

Classification of lagoons is based on degree of mechanical mixing provided.

Aerobic lagoon: The reactor is called an aerobic lagoon, when sufficient energy is supplied to keep the entire contents, including the sewage solids, mixed and aerated. To meet suspended- solids effluent standards, solids are removed from the effluent coming from an aerobic lagoon.

Facultative lagoon: In facultative lagoon, only enough energy is supplied to mix the liquid portion of the lagoon, solids settle to the bottom in areas of low velocity gradients and proceed to degrade anaerobically and this process is different from facultative pond only in the method by which oxygen is supplied. Facultative lagoons are assumed to be completely mixed reactors without biomass recycle.

Aerobic lagoons with solid recycle: The aerobic lagoon with solids recycle is same as extended aeration activated-sludge process, but an earthen (typically lined) basin is used in place of a reinforced-concrete reactor basin. It is necessary that the aeration requirement for an aerobic lagoon with recycle must be higher than the values for an aerobic flow-through lagoon to maintain the solids in suspension.

Design of Lagoons

Process design considerations for flow-through lagoons

- BOD removal

- Effluent characteristics

- Temperature effect

- Oxygen requirement

- Energy requirement for mixing

- Solids separation

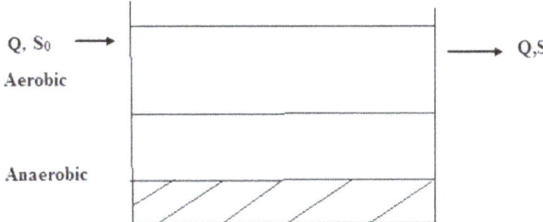

Applying mass balance on lagoon given in above figure

$$BOD_{in} = BOD_{out} + BOD_{consumed}$$

$$QS_0 = QS + V(kS)$$

$$\frac{S}{S_0} = \frac{1}{1 + k(V/Q)} = \frac{1}{1 + k\theta}$$

Where, S/S_0=fraction of soluble BOD remaining, k=reaction rate coefficient (d^{-1}), θ=hydraulic detention time (d^{-1}), V= reactor volume (m^3), and Q= flow rate (m^3/d).

If several reactors are arranged in series, the effluent of one pond becomes the influent to the next. A substrate balance written across a series of n reactors results in following equation:

$$\frac{S_n}{S_0} = \frac{1}{\left(1+\left(k\theta/n\right)\right)^n}$$

A wide range of values for k is available in the literature. Although many variables relating to both the reactor and wastewater affect the value of k, water temperature affects it most significantly. k value at any temperature can be find out by following equation:

$$k_T = k_{20}\varphi^{T-20}$$

Where, k_{20} = reaction rate constant at 20°C (ranges from 0.2 to 1.0) and φ =temperature coefficient ranges from 1.03 to 1.12.

Problem: Wastewater flow from a small community averages 3400 m^3/d during the winter and 6600 m^3/d during the summer. The average temperature of the coldest month is 10°C, and the average temperature of the warmest month is 30°C. The average BOD$_5$ is 200 mg/L with 70% being soluble. The reaction coefficient k is 0.23 d^{-1} at 20°C, and the value of temperature coefficient is 1.06. Prepare a preliminary design for a facultative pond treatment system for the community to remove 90% of the soluble BOD.

    a) Find volume of facultative lagoon to remove 90% of the soluble of BOD.

    b) Find the dimensions of three square lagoons in series with depth 1.5 m.

Solution:

    (a) Estimation of rate constants at given temperature

        Summer: $k_{25} = 0.23(1.06)^{30-20} = 0.411\,\mathrm{d}^{-1}$

        Winter: $k_{10} = 0.23(1.06)^{10-20} = 0.128\,\mathrm{d}^{-1}$

    (b) Estimation of volume of lagoon

        Summer: $\dfrac{S}{S_0} = \dfrac{1}{1+k(V/Q)} \Rightarrow \dfrac{20}{200} = \dfrac{1}{1+0.411(V/6600)}$

                $V = 144525.5\,\mathrm{m}^3$

        Winter: $\dfrac{20}{200} = \dfrac{1}{1+0.128(V/3400)}$

                $V = 239062\,\mathrm{m}^3$

(c) Estimation of dimensions of three square lagoons in series

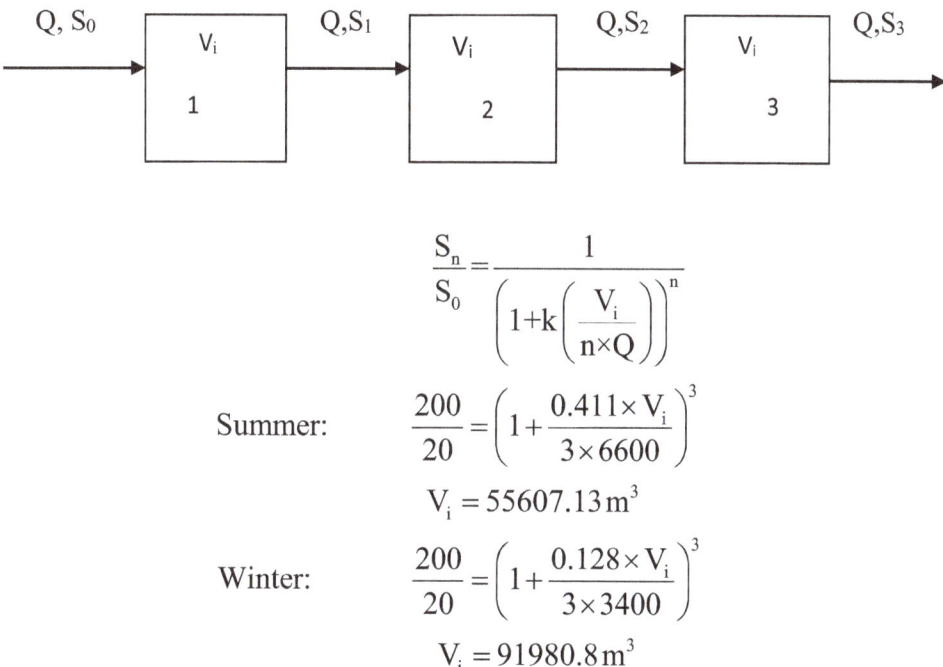

$$\frac{S_n}{S_0} = \frac{1}{\left(1+k\left(\dfrac{V_i}{n \times Q}\right)\right)^n}$$

Summer: $\dfrac{200}{20} = \left(1 + \dfrac{0.411 \times V_i}{3 \times 6600}\right)^3$

$V_i = 55607.13\, m^3$

Winter: $\dfrac{200}{20} = \left(1 + \dfrac{0.128 \times V_i}{3 \times 3400}\right)^3$

$V_i = 91980.8\, m^3$

Trickling Filter

A trickling filter plant in the United Kingdom: The effluent from the primary settling tanks is sprayed onto a bed of coarse gravel (Benfleet Sewage Treatment Plant)

A trickling filter is a type of wastewater treatment system first used by Dibden and Clowes It consists of a fixed bed of rocks, lava, coke, gravel, slag, polyurethane foam, sphagnum peat moss, ceramic, or plastic media over which sewage or other wastewater flows downward and causes a layer of microbial slime (biofilm) to grow, covering the bed of media. Aerobic conditions are maintained by splashing, diffusion, and either by forced-air flowing through the bed or natural convection of air if the filter medium is porous.

The terms trickle filter, trickling biofilter, biofilter, biological filter and biological trickling filter are often used to refer to a trickling filter. These systems have also been described as roughing filters, intermittent filters, packed media bed filters, alternative septic systems, percolating filters, attached growth processes, and fixed film processes.

Process Description

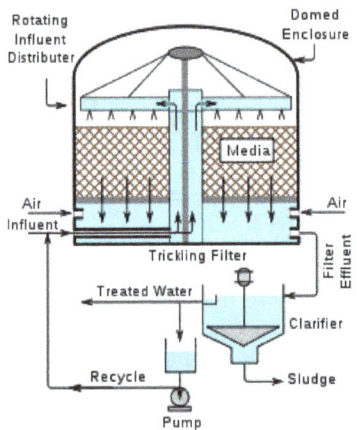

A typical complete trickling filter system

Typically, sewage flow enters at a high level and flows through the primary settlement tank. The supernatant from the tank flows into a dosing device, often a tipping bucket which delivers flow to the arms of the filter. The flush of water flows through the arms and exits through a series of holes pointing at an angle downwards. This propels the arms around distributing the liquid evenly over the surface of the filter media. Most are uncovered (unlike the accompanying diagram) and are freely ventilated to the atmosphere.

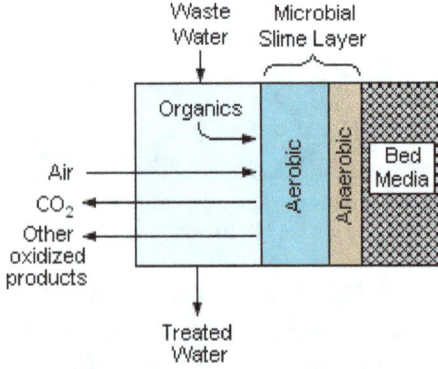

Image 1. A schematic cross-section of the contact face of the bed of media in a trickling filter

Broken trickling filter unit at the sewage treatment plant in Norton, Zimbabwe, showing importance of maintenance to prevent structural failure

The removal of pollutants from the waste water stream involves both absorption and adsorption of organic compounds and some inorganic species such as nitrite and nitrate ions by the layer of microbial bio film. The filter media is typically chosen to provide a very high surface area to volume. Typical materials are often porous and have considerable internal surface area in addition to the external surface of the medium. Passage of the waste water over the media provides dissolved oxygen which the bio-film layer requires for the biochemical oxidation of the organic compounds and releases carbon dioxide gas, water and other oxidized end products. As the bio film layer thickens, it eventually sloughs off into the liquid flow and subsequently forms part of the secondary sludge. Typically, a trickling filter is followed by a clarifier or sedimentation tank for the separation and removal of the sloughed film. Other filters utilizing higher-density media such as sand, foam and peat moss do not produce a sludge that must be removed, but require forced air blowers and backwashing or an enclosed anaerobic environment.

Biofilm

The bio-film that develops in a trickling filter may become several millimetres thick and is typically a gelatinous matrix that contains many species of bacteria, cilliates and amoeboid protozoa, annelids, round worms and insect larvae and many other micro fauna. This is very different from many other bio-films which may be less than 1 mm thick. Within the thickness of the biofilm both aerobic and anaerobic zones can exist supporting both oxidative and reductive biological processes. At certain times of year, especially in the spring, rapid growth of organisms in the film may cause the film to be too thick and it may slough off in patches leading to the "spring slough".

Design Considerations

A typical trickling filter is circular and between 10 metres and 20 metres across and between 2 metres to 3 metres deep. A circular wall, often of brick, contains a bed of filter media which in turn rests on a base of under-drains. These under-drains function both to remove liquid passing through the filter media but also to allow the free passage of air up through the filter media. Mounted in the center over the top of the filter media is a spindle supporting two or more horizontal perforated pipes which extend to the edge of the media. The perforations on the pipes are designed to allow an even flow of liquid over the whole area of the media and are also angled so that when liquid flows from the pipes the whole assembly rotates around the central spindle. Settled sewage is delivered to a reservoir at the centre of the spindle via some form of dosing mechanism, often a tipping bucket device on small filters.

Larger filters may be rectangular and the distribution arms may be driven by hydraulic or electrical systems.

Types

Single trickling filters may be used for the treatment of small residential septic tank discharges and very small rural sewage treatment systems. Larger centralized sewage treatment plants typically use many trickling filters in parallel.

Systems can be configured for single-pass use where the treated water is applied to the trickling filter once before being disposed of, or for multi-pass use where a portion of the treated water is cycled back and re-treated via a closed loop. Multi-pass systems result in higher treatment quality and assist in removing Total Nitrogen (TN) levels by promoting nitrification in the aerobic media bed and denitrification in the anaerobic septic tank. Some systems use the filters in two banks operated in series so that the wastewater has two passes through a filter with a sedimentation stage between the two passes. Every few days the filters are switched round to balance the load. This method of treatment can improve nitrification and de-nitrification since much of the carbonaceous oxidative material is removed on the first pass through the filters.

Media Types

Trickling may have a variety of types of filter media used to support the biofilm. Types of media most commonly used include coke, pumice, plastic matrix material, open-cell polyurethane foam, clinker, gravel, sand and geotextiles. Ideal filter medium optimizes surface area for microbial attachment, wastewater retention time, allows air flow, resists plugging is mechanically robust in all weathers allowing walking access across the filter and does not degrade. Some residential systems require forced aeration units which will increase maintenance and operational costs.

Industrial Wastewater Treatment

The treatment of industrial wastewater may involve specialised tricking filters which use plastic media and high flow rates. Wastewaters from a variety of industrial processes have been treated in trickling filters. Such industrial wastewater trickling filters consist of two types:

- Large tanks or concrete enclosures filled with plastic packing or other media.

- Vertical towers filled with plastic packing or other media.

The availability of inexpensive plastic tower packings has led to their use as trickling filter beds in tall towers, some as high as 20 meters. As early as the 1960s, such towers were in use at: the Great Northern Oil's Pine Bend Refinery in Minnesota; the Cities Service Oil Company Trafalgar Refinery in Oakville, Ontario and at a kraft paper mill.

The treated water effluent from industrial wastewater trickling filters is typically processed in a clarifier to remove the sludge that sloughs off the microbial slime layer attached to the trickling filter media as for other trickling filter applications.

Some of the latest trickle filter technology involves aerated biofilters of plastic media in vessels using blowers to inject air at the bottom of the vessels, with either downflow or upflow of the wastewater.

Factors Affecting the Operation of Trickling Filter

[A] Organic loading

- A high organic loading rate results in a rapid growth of biomass.

- Excessive growth may result in plugging of pores and subsequent flooding of portions of the medium.

[B] Hydraulic flow rates

- Increasing the hydraulic loading rate increases sloughing and helps to keep the bed open. Range of hydraulic and organic loading rates for trickling filters are shown in table.

[C] Relative temperature of wastewater and ambient air

- Cool water absorbs heat from air, and the cooled air falls towards toward the bottom of the filter in a concurrent fashion with the water.

- Warm water heats the air, causing it to rise through the underdrain and up through the medium.

- At temperature differentials of less than about 3 to 4°C, relatively little air movement results, and stagnant conditions prevent good ventilation.

- Extreme cold may result in icing and destruction of the biofilms.

Design Equations for Trickling Filter

[A] Tentative method of ten states of USA

The equation is given as follows:

$$E = \frac{(R/Q)+1}{(R/Q)+1.5}$$

where, Q is the flow rate, R is the recycle flow rate and E is the efficiency.

a) Loading rate

(Raw settled domestic sludge)$< 102\,\text{kg BOD}/(\text{dm}^3)$

b) R/Q should be such that

BOD entering filter (including recirculation) $\leq 3 \times$ BOD expected in effluents

[B] Velz equation

The following equation is used for a single-stage system and in the first stage of a two-stage system:

$$S_{e1} = \left[(S_i + r_1 S_{e1})/(1+r_1)\right]\exp\left[(-kDA^n / Q^n)(1.035^{T-20})\right]$$

The following equation is used for the second stage of a two-stage system:

$$S_{e2} = \left[(S_e + r_2 S_{e2})/(1+r_2)\right]\exp\left[(-kDA^n S_{e1}/Q^n S_i)(1.035^{T-20})\right]$$

Where, S_e is the effluent BOD from the filter (mg/l), S_i is the influent BOD (mg/l), r is the ratio of recirculated flow to wastewater flow, D is the filter depth (m), A is the filter plan area (m²), Q is the wastewater flow (m³/min), T is the wastewater temperature (°C), k and n are empirical coefficients (for municipal wastewaters, k = 0.02 and n = 0.5) and subscript i (i=1,2) repressent the stage number.

[C] NRC equations

The following equation is used for a single-stage system and the first stage of a two-stage system:

$$1 - (S_{e1}/S_{i1}) = 1/\left[1 + 0.532(QS_i/V_1 F_1)^{0.5}\right]$$

$$F_1 = \left[(1+r_1)/(1+0.1r_1)^2\right]$$

The following equation is used for the second stage of a two-stage system:

$$1 - (S_{e2}/S_{e1}) = 1/\left[1 + 0.532(QS_{e1}/V_2 F_2)^{0.5}\right]$$

$$F_2 = \left[(1+r_2)/(1+0.1r_2)^2\right]$$

Where, V is the filter volume (m³) and F is the recirculation factor.

[D] Eckenfelder equation (Plastic media)

The Eckenfelder equation used for plastic media is as follows:

$$S_e / S_i = \exp\left[-KD\left(QS_a^m / A\right)^{-n}\right]$$

Where, K is the observed rate constant for a given filter depth (m/d), S_a is the specific surface area of the filter (m²/m³), D is the filter depth (m), Q is the wastewater flow rate (m³/d), A is the filter plan area (ft²), and m and n are empirical coefficients.

[E] Germain/Schultz equations (Plastic media)

The Germain/Schultz equations used for plastic media are as follows:

$$S_e / S_i = \exp\left[-K_{20,i}D_i\left(Q / A\right)^{-n}\right]$$

$$k_{20,2} = k_{20,1}\left(D_1 / D_2\right)^x$$

Where, $k_{20,I}$ is the treatability constant corresponding to a specific filter depth D_i at 200C, (m³/min)ⁿ m, Q is the wastewater flow (m³/min), n and x are empirical constants (n is usually 0.5; x is 0.5 for rock and 0.3 for cross-flow plastic media).

Problem: Calculate the values of k_f and influent BOD (S_o) to trickling filter for R/Q (ratio of recycle flow rate to hydraulic loading) value of 1.65. Given that: raw settled BOD after primary settling (S_a)=220 mg/l; hydraulic loading (Q)=30 m³/(d. m²); depth of filter (D)=1.5 m; n=0.5 and effluent BOD after secondary settling (S)=35 mg/l.

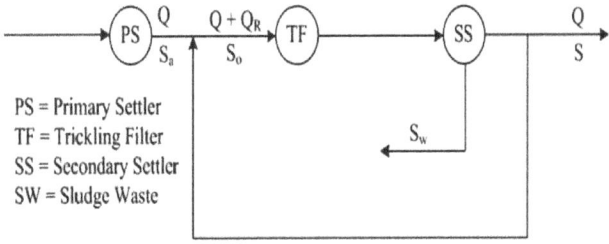

PS = Primary Settler
TF = Trickling Filter
SS = Secondary Settler
SW = Sludge Waste

Solution:

$$QS_a + Q_R S = \left(Q + Q_R\right)S_o$$

$$S = \frac{\left(Q + Q_R\right)S_o - QS_a}{Q_R} = \frac{\left(1 + \left(Q_R / Q\right)\right)S_o - S_a}{\left(Q_R / Q\right)}$$

Also,

$$S_o = \frac{QS_a + Q_R S}{Q + Q_R} = \frac{S_a + \left(Q_R / Q\right)S}{1 + \left(Q_R / Q\right)}$$

On putting values in last equation, we get,

$$S_o = \frac{220 + 1.63 \times 35}{1 + 1.65} = 104.81 \, \text{mg/l}$$

From Eckenfelder equation,

$$\frac{S}{S_0} = \exp\left[-\frac{k_f D}{Q^n}\right]$$

$$\frac{35}{104.81} = \exp\left[-\frac{k_f \times 1.5}{30^{0.5}}\right]$$

$$k_f = 4.004 \, \text{m}^{-1/2} \text{d}^{-1/2}$$

Sequencing Batch Reactor

Sequencing batch reactors (SBR) or sequential batch reactors are a type of activated sludge process for the treatment of wastewater. SBR reactors treat wastewater such as sewage or output from anaerobic digesters or mechanical biological treatment facilities in batches. Oxygen is bubbled through the mixture of wastewater and activated sludge to reduce the organic matter (measured as biochemical oxygen demand (BOD) and chemical oxygen demand (COD)). The treated effluent may be suitable for discharge to surface waters or possibly for use on land.

While there are several configurations of SBRs, the basic process is similar. The installation consists of one or more tanks that can be operated as plug flow or completely mixed reactors. The tanks have a "flow through" system, with raw wastewater (*influent*) coming in at one end and treated water (*effluent*) flowing out the other. In systems with multiple tanks, while one tank is in settle/decant mode the other is aerating and filling. In some systems, tanks contain a section known as the bio-selector, which consists of a series of walls or baffles which direct the flow either from side to side of the tank or under and over consecutive baffles. This helps to mix the incoming Influent and the *returned activated sludge* (RAS), beginning the biological digestion process before the liquor enters the main part of the tank.

Treatment Stages

There are five stages in the treatment process:

1. Fill

2. React

3. Settle

4. Decant

5. Idle

The inlet valve opens and the tank is being filled in, while mixing is provided by mechanical means (no air). This stage is also called the anoxic stage.

Aeration of the mixed liquor is performed during the second stage by the use of fixed or floating mechanical pumps or by transferring air into fine bubble diffusers fixed to the floor of the tank.

No aeration or mixing is provided in the third stage and the settling of suspended solids starts.

During the fourth stage the outlet valve opens and the "clean" supernatant liquor exits the tank.

Removal of Constituents

Aeration times vary according to the plant size and the composition/quantity of the incoming liquor, but are typically 60 to 90 minutes. The addition of oxygen to the liquor encourages the multiplication of aerobic bacteria and they consume the nutrients. This process encourages the conversion of nitrogen from its reduced ammonia form to oxidized nitrite and nitrate forms, a process known as nitrification.

To remove phosphorus compounds from the liquor, aluminium sulfate (alum) is often added during this period. It reacts to form non-soluble compounds, which settle into the sludge in the next stage.

The *settling* stage is usually the same length in time as the aeration. During this stage the sludge formed by the bacteria is allowed to settle to the bottom of the tank. The aerobic bacteria continue to multiply until the dissolved oxygen is all but used up. Conditions in the tank, especially near the bottom are now more suitable for the anaerobic bacteria to flourish. Many of these, and some of the bacteria which would prefer an oxygen environment, now start to use oxidized nitrogen instead of oxygen gas (as an alternate terminal electron acceptor) and convert the nitrogen to a gaseous state, as nitrogen oxides or, ideally, molecular nitrogen (dinitrogen, N_2) gas. This is known as denitrification.

Anoxic SBR can be used for anaerobic processes, such as the removal of ammonia via Anammox, or the study of slow-growing microorganisms. In this case, the reactors are purged of oxygen by flushing with inert gas and there is no aeration.

As the bacteria multiply and die, the sludge within the tank increases over time and a waste activated sludge (WAS) pump removes some of the sludge during the settle stage to a digester for further treatment. The quantity or "age" of sludge within the tank is closely monitored, as this can have a marked effect on the treatment process.

The sludge is allowed to settle until clear water is on the top 20 to 30 percent of the tank contents.

The decanting stage most commonly involves the slow lowering of a scoop or "trough" into the basin. This has a piped connection to a lagoon where the final effluent is stored for disposal to a wetland, tree growing lot, ocean outfall, or to be further treated for use on parks, golf courses etc.

Conversion

In some situations in which a traditional treatment plant cannot fulfill required treatment (due to higher loading rates, stringent treatment requirements, etc.) the owner might opt to convert their traditional system into a multi-SBR plant. Conversion to SBR will create a longer sludge age, minimizing sludge handling requirements downstream of the SBR.

The reverse can also be done where in SBR Systems would be converted into extended aeration (EA) systems. SBR treatment systems that could not cope up with a sudden constant increase of influent would easily be converted into EA plants. Extended aeration plants are more flexible in flow rate, eliminating restrictions presented by pumps located throughout the SBR systems. Clarifiers can be retrofitted in the equalization tanks of the SBR.

Operating Parameters in SBR Process

The treatment efficiency of SBR depends on the operating parameters such as phase duration, hydraulic retention time (HRT) and organic loading, Sludge retention time (SRT), temperature, mixed liquor suspended solids (MLSS), mixed liquor volatile suspended solids (MLVSS), dissolved oxygen (DO) concentration and the strength of wastewater.

Cycle time: A cycle in SBR comprises of fill, react, settle, decant and idle phase. The total cycle time (t_C) is the sum of all these phases.

$$t_C = t_F + t_R + t_S + t_D + t_I$$

Where, t_F is the fill time (h), t_R is the react time (h), t_S is the settle time (h), t_D is the decant time (h), and t_I is the idle time (h).

Moreover during the react phase, organic matters, nitrogen or phosphorus removal may be achieved by arresting aerobic, anoxic or anaerobic condition, respectively.

Therefore, aerobic, anoxic or anaerobic time can be found in react time (t_R). Hence

$$t_R = t_{AE} + t_{AX} + t_{AN}$$

Where, t_{AE} is the aerobic react time (h), t_{AX} is the anoxic react time (h), and t_{AN} is the anaerobic react time (h).

Volume exchange ratio (VER) and hydraulic retention time (HRT): Due to filling and decanting phase during a cycle, SBR operate with varying volume. Volume exchange ratio (VER) for a cycle is defined as V_F/V_T, Where, V_F is the filled volume of wastewater and decanted effluent for a cycle and V_T is the total working volume of the reactor.

HRT for the continuous system is defined as

$$HRT = \frac{(V_T)}{Q}$$

Where, Q is the daily waste water flow rate.

For SBR systems;

$$Q = V_F N_C$$

Where, N_C is the number of cycles per day and defined as:

$$N_C = \frac{24}{t_C}$$

Therefore, HRT for the SBR systems may be given as:

$$HRT = \frac{(t_C)}{V_F / V_T} \frac{1}{24}$$

Solid Retention Time (SRT): In biological treatment of wastewater, excess sludge is withdrawn from the reactor to control the sludge age (SRT). SRT determines the time (d) for which the biomass is retained in the reactor.

$$SRT = \frac{V_T X t_C}{V_W X_W 24}$$

Where, X is the MLSS in the reactor with full filled (mg/l), X_W is the MLSS in waste stream (mg/l), and V_W is the waste sludge volume (l).

Nitrification and Denitrification

Nitrogen is the main source of eutrophication. In this regard, the complete oxidation of nitrogen during the treatment is favorable. Biological nitrogen is removed in two stages: aerobic nitrification and anoxic denitrification. In the nitrification process, ammonia $\left(N-NH_4^+\right)$ is oxidized to nitrite $\left(N-NO_2^-\right)$ (equation 3.4.8) by autotrophic bacteria called Nitroso-bacteria and generated nitrite is oxidized to nitrate $\left(N-NO_3^-\right)$ (equation 3.4.9) by another group of autotrophic bacteria called Nitro-bacteria under aerobic conditions and using oxygen as the electron acceptor.

$$2NH_4^+ + 3O_2 \rightarrow 2NO_2^- + 2H_2O + 4H^+$$
$$2NO_2^- + O_2 \rightarrow 2NO_3^-$$

The autotrophic bacteria produce energy for their multiplication from the oxidation of inorganic nitrogen compounds, using inorganic carbon as their source of cellular carbon. During the nitrification, alkalinity of wastewater is used which reduces the pH of influent wastewater and required amount of alkalinity to carry out the reaction (equation 3.4.8, 3.4.9) in the $CaCO_3$ form, can be calculated by the following equation;

$$NH_4^+ + 2HCO_3^- + 2O_2 \rightarrow NO_3^- + 3H_2O + 2CO_2$$

Biological denitrification involves the biological oxidation of many organic substrates in wastewater treatment using nitrate or nitrite as the electron acceptor under the anoxic condition or limited dissolved oxygen (DO) concentrations and nitrate is degraded to nitric oxide, nitrous oxide, and nitrogen gas [4-6] by following any of the two different routes. One of these routes predominates depending on the dissolved oxygen concentration.

$$NH_4^+ \rightarrow NO_2^- \rightarrow NO_3^- \rightarrow NO_2^- \rightarrow NO \rightarrow N_2O \rightarrow N_2O$$
or
$$NH_4^+ \rightarrow NO_2^- \rightarrow NO \rightarrow N_2O \rightarrow N_2$$

During the denitrification process, pH of influent wastewater increases because of increase of alkalinity. Both heterotrophic and autotrophic bacteria are capable of denitrification. Most of these heterotrophic bacteria are facultative aerobic organisms with the ability to use oxygen as well as nitrate or nitrite, and some can also carry out fermentation in the absence of nitrate or oxygen.

Advantages and Disadvantages of SBR

Advantages

- Equalization, primary clarification (in most cases), biological treatment, and secondary clarification can be achieved in a single reactor vessel.

- Operating flexibility and control.

- Potential capital cost savings by eliminating clarifiers and other equipments.

Disadvantages

- A higher level of sophistication, (compared to conventional systems), especially for larger systems, of timing units and controls is required.

- Higher level of maintenance (compared to conventional systems) associated with more sophisticated controls, automated switches and automated valves.

- Potential of discharging floating or settled sludge during the draw or decant phases with some SBR configurations.

- Potential plugging of aeration devices during selected operating cycles, depending on the aeration system used by the manufacturer.

Upflow Anaerobic Sludge Blanket Digestion

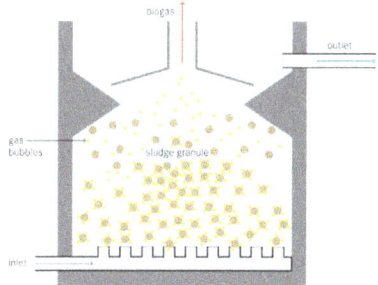

Schematic of an upflow anaerobic sludge blanket reactor (UASB): Wastewater enters the reactor from the bottom and flows upward

Upflow anaerobic sludge blanket (UASB) technology, normally referred to as UASB reactor, is a form of anaerobic digester that is used for wastewater treatment.

The UASB reactor is a methanogenic (methane-producing) digester that evolved from the anaerobic clarigester. A similar but variant technology to UASB is the expanded granular sludge bed (EGSB) digester.

Process Description

UASB uses an anaerobic process whilst forming a blanket of granular sludge which suspends in the tank. Wastewater flows upwards through the blanket and is processed (degraded) by the anaerobic microorganisms. The upward flow combined with the settling action of gravity suspends the blanket with the aid of flocculants. The blanket begins to reach maturity at around three months. Small sludge granules begin to form whose surface area is covered in aggregations of bacteria. In the absence of any support matrix, the flow conditions create a selective environment in which only those microorganisms capable of attaching to each other survive and proliferate. Eventually the aggregates form into dense compact biofilms referred to as "granules".

Biogas with a high concentration of methane is produced as a by-product, and this may be captured and used as an energy source, to generate electricity for export and to cover its own running power. The technology needs constant monitoring when put into use to ensure that the sludge blanket is maintained, and not washed out (thereby losing the effect). The heat produced as a by-product of electricity generation can be reused to heat the digestion tanks.

The blanketing of the sludge enables a dual solid and hydraulic (liquid) retention time in the digesters. Solids requiring a high degree of digestion can remain in the reactors for periods up to 90 days. Sugars dissolved in the liquid waste stream can be converted into gas quickly in the liquid phase which can exit the system in less than a day.

UASB reactors are typically suited to dilute waste water streams (3% TSS with particle size >0.75mm).

Design

UASB reactor shown is the larger tank. Hiriya, Tel Aviv, Israel

With UASB, the process of settlement and digestion occurs in one or more large tank(s). The effluent from the UASB, which has a much reduced biochemical oxygen demand (BOD) concentration, usually needs to be treated further, for example with the activated sludge process, depending on the effluent quality requirements.

Sludge Separation and Drying

Sludge

- The polluted solid-liquid matter that is skimmed off or removed from wastewater during primary, secondary and tertiary treatment.

- It contains 0.25 to 12% organic to inorganic solid content

- Constituents

    - Organic material, nutrients, pathogens, metals, toxic substances

Goals of Sludge Management

- Stabilize sludge

- Kill pathogens

- Decrease water content from 0.5-2% solids to 6 to 12% solids

Sludge Processing

(a) Thickening

(b) Conditioning, Stabilization, Disinfection

(c) Dewatering

(d) Drying

(e) Composting

(f) Incineration

(g) Final Disposal

[A] Sludge Thickening

- Thickening: Capacity of sludge to increase concentration of solid in sludge

- Purpose: To decrease volume

- Benefits:

    - Reduces required capacity of downstream equipment

    - Reduce chemicals for conditioning

    - Reduce heat required by digesters

    - Reduce volume for transportation

- Equipment types

    - Gravity

    - Gravity Belt Thickener (GBT)

    - Flotation

- Rotary drum

- Centrifuge

[B] Sludge Conditioning

- Sludge particles are negative (anionic) in surface charge

- The negative surface charge leads to electrostatic repulsive forces which hamper the settling process of the sludge particles.

- Cationic conditioning agents minimizes the electrostatic repulsive force and starts floc formation

- Chemical conditioning is similar to flocculation/coagulation process

[C] Sludge Dewatering

- Mostly done in filtration type of units where solid particles from a fluid are retained on a filtering medium which allows the water to pass through it.

- Five types of equipment

  - Belt Filter Press (18-25%)

  - Centrifuge (30-35%)

  - Recessed Chamber Press

  - Vacuum Filtration

  - Drying Beds

[D] Sludge Drying

- Direct: Sludge in contact with heat surface, e.g. fluidized bed dryer, revolving drum dryers

- Indirect: There is no direct contact between heat source and sludge, e.g. Disc dryer

- More expensive than mechanical methods such as pressing or centrifugation

- Yields greater volume reduction and a storable free flowing and hygienic product.

- End product can be used as

  - fertilizer/soil conditioner in agriculture and forestry

  - fuel in cement kilns, power plants and incinerators

  - top soil, landscaping, and landfilling use.

[E] Sludge Composting

- Can be applied to either digested or non-digested sludge

- Need to have sufficient mixture of organic matter content and water

- Carbon to nitrogen ratio: 25-30

- May be used as pretreatment to incineration

- Advantages

  - reduction in volume of materials to be transported for distribution in agricultural fields

  - allows the facilitation of storage

  - easier to spread

  - control in the nutrients in the compost

  - compost is more hygienic than raw sludge application

- Disadvantage

  - costly

  - requires aeration

  - requires a market

[F] Sludge Incineration

- A method used for drying and reducing sludge volume and weight. Since incineration requires auxiliary fuel to obtain and maintain high temperature and to evaporate the water contained in the incoming sludge, concentration techniques should be applied before incineration.

- Sludge incineration is a two-step process involving drying and combustion after a preceding dewatering process, such as filters, drying beds, or centrifuges.

- Multiple Hearths

  - Top – Drying

  - Middle – Incineration

  - Lower – Cooling

- Flue gas – need to be treated

[G] Sludge Disposal

- Agriculture: For raw and treated sludge

- Things to consider:

  - Heavy Metal content

  - Dry solid content

- Advantage:

  - Utilization of nutrients in soil (organics, nitrogen, phosphorus)

  - Cheaper (raw sludge)

- Disadvantage: need for storage facility (investment)

  - Landfilling

Dewatering Filters

- Filtration is the removal of solid particles from a fluid by passing the fluid through a filtering medium, or septum, on which the solids are deposited. However, the mechanical separation (filtration or clarification) of primary sludge is only partially effective as a treatment because 30 to 40 % of BOD and COD are water soluble and cannot be so removed.

- Filtration is generally complete in 1 to 2 days and results in solids concentration as high as 15 to 20%. The rate of filtration depends drainability of the sludge, which in turn is related to the specific resistance

Types of Dewatering Filters

[A] Rotary Drum Vacuum Filters (RDVF)

- The filtration, washing, partial drying and discharge of the sludge all take place simultaneously.

- Process involves sucking of liquid through a moving septum to deposit a cake of solids.

- The cake is moved out of the filtering zone, washed, sucked dry, and dislodged from the septum, which then reenters the slurry to pick up another load of solids.

Table: Advantages and disadvantages of rotary drum filters

Advantages	Disadvantages
Filter is entirely automatic.	Maximum available pressure difference is limited as it being a vacuum filter.
Large capacity, hence large quantities can be filtered.	Difficulty in filtration of hot liquids because of their tendency to boil.

Cakes of varying thickness can be built by varying speed which results in removal of fine or coarser solids easily.	Initial cost of filter and vacuum equipment is high.
Low maintenance cost.	These are inflexible and do not perform well if their feed stream conditions are changing.

[B] Filter press

- It contains a set of plates designed to provide a series of chambers or compartments in which solids may collect.

- The plates are covered with a filter medium such as canvas.

- Slurry is admitted to each compartment under pressure; liquor passes through the canvas and out a discharge pipe, leaving a wet cake of solids behind.

- During operation, when the frames are full of solids and no more slurry can enter. The press is then said to be jammed.

- Wash liquid may then admitted to remove soluble impurities from the solids.

[C] Horizontal belt filter

- It is suitable for coarser particles as compared to rotary-drum filters.

- Feed slurry flows onto the belt from a distributor at one end of the unit; filtered and washed cake is discharged from the other.

- It is suitable for waste treatment as it is available in various sizes. They are available in sizes ranging from 0.6 to 5.5 m wide and 4.9 to 33.5 m long, with filtration areas up to 110 m².

[D] Rotating-leaf filter

- During filtration, the slurry enters, the filtrate exits, and solids are retained on leaves and covered with a filter cloth.

- Upon completion of filtration, the washing and drying bottom closure opens.

- The drive motor starts and rotates the stack of filter leaves.

- Centrifugal force causes the solids to move off the filter leaves, strike the inside wall of the tank and flow down to solid exit.

- Sizes are available up to 540 ft² per unit.

[E] Deep bed filter

- Filters with deep beds of sand, diatomaceous earth, coke, charcoal, and other inexpensive packing materials are normally used.

- Without preseparation the bed becomes loaded quickly.

- When the particle and bacteria in sizes smaller than the interstices of the bed, plus suspended BOD, are remove from the liquid, exceptional clarity is obtained.

- The dissolved substances, including dissolved BOD are not removed.

Thermal Dryers

Heat treatment followed by filtration is economical for dewatering sludge without using chemicals. Thermal drying of the sludge is economical only if a market for the product is available. Several types of thermal dryers used by the chemical process industry can be applied to sludge drying. The sludge is always dewatered prior to drying, regardless of the type of dryer selected.

Types of Dryers

[A] Flash dryer

- It operates by promoting contact between the wet sludge and a hot gas stream.

- Drying takes place in less than 10 sec of violent action, either in a vertical tube or in a cage mill.

- A cyclone, with a bag filter or wet scrubber, if necessary, separates the solid from the gas phase.

- The vapors are returned through preheaters to the furnace, minimizing odor problems.

- A portion of the solid product is often returned to precondition the wet sludge.

- Being of only moderate thermal efficiency, this type of furnace is appropriate only for low sludge flows and where heat is available cheaply.

[B] Screw conveyor dryers

- It uses a hollow shaft and blades through which hot gas or water is pumped.

- The heat is transferred to the sludge as it is conveyed through the dryer.

[C] Multiple-hearth dryer

- These are converted multiple-hearth furnaces.

- The wet sludge can be mixed with dry product as it descends through the furnace.

- Fuel burners are located both on top and bottom.

- The outlet temperature of the gases is approximately 400 °C, while that of the wet sludge at the upper drying levels barely exceeds 70°C.

[E] Rotary dryers

- It consists of a rotating cylinder through which the sludge moves.

- Various types of blades or flights are installed in the dryers depending on the type of material being dried.

- Drying takes place by direct contact with heated air.

- With a combustion temperature of 900 to 1000°C and 50 % excess air, the outlet temperature of gases from sewage sludge is around 300°C.

[F] Atomized spray dryers

- It has been used for many years in the chemical process industry.

- Spraying solids counter-currently into a downward draft of hot gas dries although concurrent spray dryers are also used in the chemical industry.

Membrane based Technologies

Membrane

Membrane can be described as a thin layer of material that is capable of separating materials as a function of their physical and chemical properties when a driving force is applied across the membranes. Physically membrane could be solid or liquid.

In membrane separation processes, the influent to the membrane module is known as the feed stream (also known as the feed water), the liquid that passes through the semipermeable membrane is known as permeate (also known as the product stream or permeating stream) and the liquid containing the retained constituents is known as the concentrate also known as retained phase.

Membrane Process Classification

Membrane processes can be classified in a number of different ways :

- The type of material from which the membrane is made

- The nature of the driving force

- The separation mechanism

- The nominal size of the separation achieved

Table: General characteristics of membrane processes

Membrane process	Driving force	Method of separation	Operating structure (pore size)	Typical operating range, µm	Permeate description	Range of application
Microfiltration	Hydrostatic pressure difference	Sieving mechanism	Macropores (>50 nm)	0.08 - 2.0	Water + dissolved solutes	Sterile filtration clarification
Ultrafiltration	Hydrostatic pressure difference	Sieving mechanism	Mesopores (2 -50 nm)	0.005 – 0.2	Water + small molecules	Separation of macromolecular solutions
Nanofiltration	Hydrostatic pressure difference	Sieving mechanism + solution/ diffu sion	Micropores (<2 nm)	0.001 – 0.01	Water + very small molecules, ionic solutes	Removal of small molecules, small harness, viruses
Reverse osmosis	Hydrostatic Pressure difference	Solution Diffusion mechanism + exclusion	Dense (<2 nm)	0.0001 –0.001	Water + small molecules	Separation of salts and microsolutes from solutions
Dialysis	Concentration gradient	Diffusion in convection free layer	Mesopores (2 -50 nm)	-	Water + ionic solutes	Separation of salts and microsolutes from macromolecular solutions
Electrodialysis	Electrical potential gradient	Electrical charge of particle and size	Micropores (<2 nm)	-		Desalting of ionic solution

Table: Advantages & disadvantages of membrane technologies

Advantages	Disadvantages
Microfiltration and ultrafiltration	
➢ Can reduce the amount of treatment chemicals	➢ Uses more electricity; high-pressure systems can be energy-intensive
➢ Smaller space requirements (footprint); membrane equipment requires 50 to 80 percent less space than conventional plants	➢ May need pretreatment to prevent fouling; pretreatment facilities increase space needs and overall costs
➢ Reduced labour requirements; can be automated easily	➢ May require residuals handling and disposal of concentrate

➤ New membrane design allows use of lower pressures; system cost may be competitive with conventional wastewater-treatment processes	➤ Require replacement of membranes about every 3 to 5 years
➤ Remove protozoan cysts, oocysts, and helminth ova; may also remove limited amounts of bacteria and viruses	➤ Scale formation can be a serious problem. Scale-forming potential difficult to predict without field testing
	➤ Flux rate (the rate of feedwater flow through the membrane) gradually declines over time. Recovery rates may be considerably less than 100 percent
	➤ Lack of a reliable low-cost method of monitoring performance
Reverse osmosis	
➤ Can remove dissolved constituents	➤ Works best on ground water or low solids surface water or pretreated wastewater effluent
➤ Can disinfect treated water	➤ Lack of a reliable low-cost method of monitoring performance
➤ Can remove NDMA and other related organic compounds	➤ May require residuals handling and disposal of concentrate
➤ Can remove natural organic matter (a disinfection by-product precursor) and inorganic matter	➤ Expensive compared to conventional treatment

Membrane Materials & Configurations

- Membranes can be made from a number of different organic and inorganic materials. The membranes used for wastewater treatment are typically organic. The principle types of membranes used include polypropylene, cellulose acetate, aromatic polyamides, and thin- film composite (TFC).

- Membranes used for the treatment of water and wastewater typically consist of a thin skin having a thickness of about 0.20 to 0.25 µm supported by a more porous structure of about 100 µm in thickness.

- Term 'module' is used to describe a complete unit comprised of the membranes, the pressure support structure for the membranes, the feed inlet and outlet permeate and retentate ports, and an overall support structure.

- The principle types of membrane modules used for wastewater treatment are 1) tubular, 2) spiral wound, 3) hollow fibre, 4) flat.

Table: Comparison of different membrane configurations

Membrane geometry	Suspended solids tolerance	Control of fouling	Cleaning easiness	Packing density	Cost for unit of volume
Tubular	Good	Excellent	Excellent	Low- medium	Medium- high

Spi-ral-wound	Low	Limited	Medium	High	Low
Hollow fibre (external feed)	Scant (good)	Scant (good)	Scant (good)	Excellent	High (low)
Flat	Medium	Good	Medium	Medium	Medium-low

Membrane Fouling

Membranes can be seen as sieves retaining part of the feed. As a consequence, deposits of the retained material will accumulate at the feed side of the membrane. In time this might hamper the selectivity and productivity of the separation process. This process is called fouling. koros et al gave the definition of fouling as "The process resulting in loss of performance of a membrane due to deposition of suspended or dissolved substances on its external surfaces, at its pore openings, or within its pores". Membrane fouling is an important consideration in the design and operation of membrane systems as it affects pretreatment needs, cleaning requirements, operating conditions, cost, and performance.

Three approaches are used to control membrane fouling:

1) Pretreatment of the feed water: pretreatment is used to reduce the TSS and bacterial content of the feed water

2) Membrane backflushing: to eliminate the accumulated material from the membrane surface with water and/or air.

3) Chemical cleaning of the membranes: Chemical treatment is used to remove constituents that are not removed during conventional backwashing. Chemical precipitates can be removed by altering the chemistry of the feed water and by chemical treatment.

Adsorption

- Adsorption can be simply defined as the concentration of a solute, which may be molecules in a gas stream or a dissolved or suspended substance in a liquid stream, on the surface of a solid.

- In an adsorption process, molecules or atoms or ions in a gas or liquid diffuse to the surface of a solid, where they bond with the solid surface or are held there by weak inter- molecular forces. The adsorbed solute is called the adsorbate, and the solid material is the adsorbent.

- Activated clays, activated carbons, fuller earths, bauxite, alumina, bone char, molecular sieves, synthetic polymeric adsorbents, silica gel, etc. are the main types of adsorbents used in the industry.

- There are basically two types of adsorption processes: one is physical adsorption (physisorption) and the second is chemisorption.

Diffusion of Adsorbate

There are essentially four stages in the adsorption of an organic/inorganic species by a porous adsorbent :

1. Transport of adsorbate from the bulk of the solution to the exterior film surrounding the adsorbent particle;

2. Movement of adsorbate across the external liquid film to the external surface sites on the adsorbent particle (film diffusion);

3. Migration of adsorbate within the pores of the adsorbent by intraparticle diffusion (pore diffusion);

4. Adsorption of adsorbate at internal surface sites.

All these processes play a role in the overall sorption within the pores of the adsorbent. In a rapidly stirred, well mixed batch adsorption, mass transport from the bulk solution to the external surface of the adsorbent is usually fast. Therefore, the resistance for the transport of the adsorbate from the bulk of the solution to the exterior film surrounding the adsorbent may be small and can be neglected. In addition, the adsorption of adsorbate at surface sites (step 4) is usually very rapid and thus offering negligible resistance in comparison to other steps, i.e. steps 2 and 3. Thus, these processes usually are not considered to be the rate-limiting steps in the sorption process.

In most cases, steps (2) and (3) may control the sorption phenomena. For the remaining two steps in the overall adsorbate transport, three distinct cases may occur:

Case I:	external transport	>	internal transport.
Case II:	external transport	<	internal transport.
Case III:	external transport	≈	internal transport.

In cases I and II, the rate is governed by film and pore diffusion, respectively. In case III, the transport of ions to the boundary may not be possible at a significant rate, thereby, leading to the formation of a liquid film with a concentration gradient surrounding the adsorbent particles.

Usually, external transport is the rate-limiting step in systems which have (a) poor phase mixing, (b) dilute concentration of adsorbate, (c) small particle size, and (d) high affinity of the adsorbate for the adsorbent. In contrast, the intra-particle step limits the overall transfer for those systems that have (a) a high concentration of adsorbate, (b) a good phase mixing, (c) large particle size of the adsorbents, and (d) low affinity of the adsorbate for adsorbent.

The possibility of intra-particle diffusion can be explored using the intra-particle diffusion model.

$$q_t = k_{id}t^{1/2} + I$$

Where, q_t is the amount of the adsorbate adsorbed on the adsorbent (mg/g) at any t and k_{id} is the intra-particle diffusion rate constant, and values of I give an idea about the thickness of the boundary layer.

In order to check whether surface diffusion controls the adsorption process, the kinetic data can be analyzed using Boyd kinetic expression which is given by :

$$F = 1 - \frac{6}{\pi^2}\exp(-B_t) \quad or \quad B_t = -0.4977 - \ln(1 - F)$$

Where, $F(t) = q_t/q_e$ is the fractional attainment of equilibrium at time t, and B_t is a mathematical function of F.

However, if the data exhibit multi-linear plots, then two or more steps influence the overall adsorption process. In general, external mass transfer is characterized by the initial solute uptake and can be calculated from the slope of plot between C/C_o versus time. The slope of these plots can be calculated either by assuming polynomial relation between C/C_o and time or it can be calculated based on the assumption that the relationship was linear for the first initial rapid phase.

Adsorption Kinetic

Pseudo-first-order and pseudo-second-order model: The adsorption of adsorbate from solution to adsorbent can be considered as a reversible process with equilibrium being established between the solution and the adsorbate. Assuming a non-dissociating molecular adsorption of adsorbate molecules on adsorbent, the sorption phenomenon can be described as the diffusion controlled process.

Using first order kinetics it can be shown that with no adsorbate initially present on the adsorbent, the uptake of the adsorbate by the adsorbent at any instant t is given as.

$$q_t = q_e\left[1 - \exp(-k_f t)\right]$$

where, q_e is the amount of the adsorbate adsorbed on the adsorbent under equilibrium condition, k_f is the pseudo-first order rate constant.

The pseudo-second-order model is represented as:

$$q_t = \frac{tk_S q_e^2}{1 + tk_S q_e}$$

The initial sorption rate, h (mg/g min), at t→0 is defined as

$$h = k_s q_e^2$$

Adsorption Isotherm

Equilibrium adsorption equations are required in the design of an adsorption system and their subsequent optimization. Therefore it is important to establish the most appropriate correlation for the equilibrium isotherm curves.

Srivastava et al. have discussed the theory associated with the most commonly used isotherm models. Various isotherms namely Freundlich, Langmuir, Redlich-Peterson (R-P) and Tempkin which are given in following table are widely used to fit the experimental data:

Table: Various isotherm equations for the adsorption process

Isotherm	Equation
Freundlich	$q_e = K_F C_e^{1/n}$
Langmuir	$q_e = \dfrac{q_m K_L C_e}{1 + K_L C_e}$
R-P	$q_e = \dfrac{K_R C_e}{1 + a_R C_e^{\beta}}$
Tempkin	$q_e = B_T \, ln(K_T C_e)$

K_R: R–P isotherm constant (l/g), a_R: R–P isotherm constant (l/mg), β: Exponent which lies between 0 and 1, C_e: Equilibrium liquid phase concentration (mg/l), K_F: Freundlich constant (l/mg), $1/n$: Heterogeneity factor, K_L: Langmuir adsorption constant (l/mg), q_m: adsorption capacity (mg/g), K_T: Equilibrium binding constant (l/mol), B_T: Heat of adsorption.

The Freundlich isotherm is derived by assuming a heterogeneous surface with a non-uniform distribution of heat of adsorption over the surface, whereas in the Langmuir theory the basic assumption is that the sorption takes place at specific homogeneous sites within the adsorbent. The R-P isotherm incorporates three parameters and can be applied either in homogenous or heterogeneous systems. Tempkin isotherm assumes that the heat of adsorption of all the molecules in the layer decreases linearly with coverage due to adsorbent-adsorbate interactions, and the adsorption is characterized by a uniform distribution of binding energies, up to some maximum binding energy.

Factors Controlling Adsorption

The amount of adsorbate adsorbed by an adsorbent from adsorbate solution is influenced by a number of factors as discussed below:

Nature of Adsorbent

The physico-chemical nature of the adsorbent is important. Adsorbents differ in their specific surface area and affinity for adsorbate. Adsorption capacity is directly proportional to the exposed surface. For the non-porous adsorbents, the adsorption capacity is inversely proportional to the particle diameter whereas for porous material it is practically independent of particle size. However, for porous substances particle size affects the rate of adsorption. For substances like granular activated carbon, the breaking of large particles to form smaller ones open up previously sealed channels making more surface accessible to adsorbent.

Pore sizes are classified in accordance with the classification adopted by the International Union of Pure and Applied Chemistry (IUPAC) , that is, micro-pores (diameter (d) <20 Å), meso-pores (20 Å < d < 500 Å) and macro-pores (d > 500 Å). Micro-pores can be divided into ultra-micropores (d < 7 Å) and super micro-pores (7 Å < d < 20 Å).

pH of Solution

The surface charge as well as the degree of ionization is affected by the pH of the solution. Since the hydrogen and hydroxyl ions adsorbed readily on the adsorbent surface, the adsorption of other molecules and ions is affected by pH. It is a common observation that a surface adsorbs anions favorably at low pH and cations in high pH range.

Contact Time

In physical adsorption most of the adsorbate species are adsorbed within a short interval of contact time. However, strong chemical binding of adsorbate with adsorbent requires a longer contact time for the attainment of equilibrium. Available adsorption results reveal that the uptake of adsorbate species is fast at the initial stages of the contact period, and thereafter, it becomes slower near the equilibrium. In between these two stages of the uptake, the rate of adsorption is found to be nearly constant. This may be due to the fact that a large number of active surface sites are available for adsorption at initial stages and the rate of adsorption is a function of available vacant site. Concentration of available vacant sites decreases and there is repulsion between solute molecules thereby reducing the adsorption rate.

Initial Concentration of Adsorbate

A given mass of adsorbent can adsorb only a fixed amount of adsorbate. So the initial concentration of adsorbate solution is very important. The amount adsorbed decreases with increasing adsorbate concentration as the resistance to the uptake of solute from solution of adsorbate decreases with increasing solute concentration. The rate of adsorption is increased because of the increasing driving force.

Temperature

Temperature dependence of adsorption is of complex nature. Adsorption processes are generally exothermic in nature and the extent and rate of adsorption in most cases decreases with increasing temperature. This trend may be explained on the basis of rapid increase in the rate of desorption or alternatively explained on the basis of Le-Chatelier's principle.

Some of the adsorption studies show increased adsorption with an increase in temperature. This increase in adsorption is mainly due to an increase in number of adsorption sites caused by breaking of some of the internal bonds near the edge of the active surface sites of the adsorbent. Also, if the adsorption process is controlled by the diffusion process (intraparticle transport-pore diffusion), than the sorption capacity increases with an increase in temperature due to endothermicity of the diffusion process. An increase in temperature results in an increased mobility of the metal ions and a decrease in the retarding forces acting on the diffusing ions. These result in the enhancement in the sorptive capacity of the adsorbents.

Adsorption Operations

Fixed Bed Adsorbers

- These are used for the adsorption of dyes and colorants, refractory pollutants from wastewater.

- The size of the bed depends on the gas flow rate and the desired cycle time.

- The bed length usually varies from 0.3 to 1.3 m.

- The gas is fed downward through the adsorbent particles in the bed.

- Inside the bed, the adsorbent particles are placed on a screen, or performed plate.

- Upflow of feed is usually avoided because of the tendency of fluidization of the particles at high rates. When the adsorption reaches the desired value, the feed goes to the other bed through an automatic valve and the regeneration process starts.

- Regeneration process is usually carried out by steam, provided the solvent is immiscible with water. It may also be carried out by a hot or inert gas.

- The adsorption cycle usually varies from 2 to 24 h. for a large bed, the adsorption cycle is high, but the pressure drop and capital cost are also high.

- For a small bed, the pressure drop is less, but the separation is incomplete and more energy is required for regeneration.

The concentration of solute in the fluid phase and of the solid adsorbent phase change with time and with position in the fixed bed as adsorption proceeds. At the inlet to the bed, the solid is assumed to contain no solute at the start of the process. As the fluid first comes in contact with the inlet of the bed, most of the mass transfer and adsorption takes place here. As the fluid passes through the bed, the concentration of the fluid drops very rapidly with the distance in the bed and reaches zero well before the end of the bed of the reached.

After the short time, the solid near the entrance is almost saturated, and most of the mass transfer and adsorption now takes place at a point slightly farther from the inlet. Following figure shows the breakthrough concentration profile in the fluid at outlet of bed.

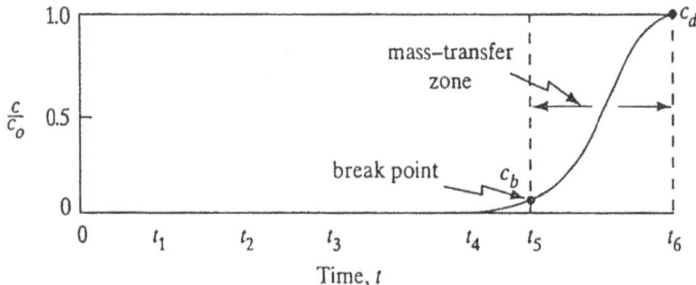

Breakthrough concentration profile in the fluid at outlet of bed.

In the figure, the outlet concentration remains near zero until the mass transfer zone starts to reach the tower outlet at time t_4. Then the outlet concentration starts to rise, and at t_5 the outlet concentration has risen to c_b, which is called the break point. The service time of the bed is given by the Bohart Adams equation.

$$t = \frac{N_o}{C_o \cdot v} \left[X - \frac{v}{K \cdot N_o} \ln\left(\frac{C_o}{C_B} - 1 \right) \right]$$

Where, t is the service time of the bed, v is the linear flow rate of the solution, X is the depth of the bed, K is the rate constant, N_o is the adsorption capacity, C_o is the concentration of the solute entering into the bed, C_B is the allowable effective concentration.

Stirred Tank Adsorbers

- Stirred tank adsorbers are generally used for the removing pollutants from the aqueous wastes.

- Such an adsorber consists of a cylindrical tank fitted with a stirrer or air sparger.

- The stirrer or air-sparger keeps the particles in the tank in suspension.

- The spent adsorbate is removed by sedimentation or filtration.

- The mode of operation may be batch or continuous.

Continuous Adsorbers

- The solid and the fluid move through the bed counter currently and come in contact with each other throughout the entire apparatus without periodic separation of the phrases.

- The solid particles are fed from the top and flow down through the adsorption and regeneration sections b gravity and are then returned to the top of the column by an air lift or mechanical conveyer.

- Multi-stage fluidized beds, in which the fluidized solids pass through down comers from stage to stage, may be used for fine particles.

References

- Wastewater engineering : treatment and reuse (4th ed.). Metcalf & Eddy, Inc., McGraw Hill, USA. 2003. p. 1456. ISBN 0-07-112250-8

- Irvine, Robert L.; Busch, Arthur W. (1979-01-01). "Sequencing Batch Biological Reactors: An Overview". Journal (Water Pollution Control Federation). 51 (2): 235–243. JSTOR 25039819

- Benidickson, Jamie (2011). The Culture of Flushing: A Social and Legal History of Sewage. UBC Press. Retrieved 2013-02-07

- Eckenfelder, Jr., W. Wesley; Cleary, Joseph G. (2014). Activated Sludge Technologies for Treating Industrial Wastewaters (1st ed.). DEStech Publications. p. 234. ISBN 978-1-60595-019-8. Retrieved 29 December 2014

- Strous, M.; Heijnen, J. J.; Kuenen, J. G.; Jetten, M. S. M. (1998). "The sequencing batch reactor as a powerful tool for the study of slowly growing anaerobic ammonium-oxidizing microorganisms". Applied Microbiology and Biotechnology. 50 (5): 589–596. doi:10.1007/s002530051340

- Tilley, E., Ulrich, L., Lüthi, C., Reymond, Ph., Zurbrügg, C. (2014) Compendium of Sanitation Systems and Technologies - (2nd Revised Edition). Swiss Federal Institute of Aquatic Science and Technology (Eawag), Duebendorf, Switzerland. ISBN 978-3-906484-57-0

4

Air Pollution: Detection and Control

The presence of particles in the air, which can be harmful for living organisms, is referred to as air pollution. It can be classified into natural contaminants, gases and vapors, aerosols, etc. Air pollution can be contained through various methods and concepts such as the use of alternative fuel, combustion control, three-way catalytic convertor, etc. The aspects elucidated in this section are of vital importance, and provide a better understanding of air pollution.

Air Pollution

Air pollution from a coking oven

Air pollution occurs when harmful substances including particulates and biological molecules are introduced into Earth's atmosphere. It may cause diseases, allergies or death in humans; it may also cause harm to other living organisms such as animals and food crops, and may damage the natural or built environment. Human activity and natural processes can both generate air pollution.

Indoor air pollution and poor urban air quality are listed as two of the world's worst toxic pollution problems in the 2008 Blacksmith Institute World's Worst Polluted Places report. According to the 2014 WHO report, air pollution in 2012 caused the deaths of around 7 million people worldwide, an estimate roughly matched by the International Energy Agency.

Pollutants

An air pollutant is a substance in the air that can have adverse effects on humans and the ecosystem. The substance can be solid particles, liquid droplets, or gases. A pollutant can be of natural origin or man-made. Pollutants are classified as primary or secondary. Primary pollutants are usually produced from a process, such as ash from a volcanic eruption. Other examples include carbon monoxide gas from motor vehicle exhaust, or the sulfur dioxide released from factories. Secondary pollutants are not emitted directly. Rather, they form in the air when primary pollutants react or interact. Ground level ozone is a prominent example of a secondary pollutant. Some pollutants may be both primary and secondary: they are both emitted directly and formed from other primary pollutants.

Before flue-gas desulfurization was installed, the emissions from this power plant in New Mexico contained excessive amounts of sulfur dioxide

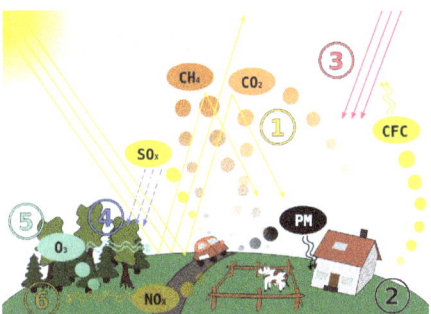

Schematic drawing, causes and effects of air pollution: (1) greenhouse effect, (2) particulate contamination, (3) increased UV radiation, (4) acid rain, (5) increased ground level ozone concentration, (6) increased levels of nitrogen oxides

Thermal oxidizers are air pollution abatement options for hazardous air pollutants (HAPs), volatile organic compounds (VOCs), and odorous emissions

Substances emitted into the atmosphere by human activity include:

- Carbon dioxide (CO_2) - Because of its role as a greenhouse gas it has been described as "the leading pollutant" and "the worst climate pollution". Carbon dioxide is a natural component of the atmosphere, essential for plant life and given off by the human respiratory system. This question of terminology has practical effects, for example as determining whether the U.S. Clean Air Act is deemed to regulate CO_2 emissions. CO_2 currently forms about 405 parts per million (ppm) of earth's atmosphere, compared to about 280 ppm in pre-industrial times, and billions of metric tons of CO_2 are emitted annually by burning of fossil fuels. CO_2 increase in earth's atmosphere has been accelerating.

- Sulfur oxides (SOx) - particularly sulfur dioxide, a chemical compound with the formula SO_2. SO_2 is produced by volcanoes and in various industrial processes. Coal and petroleum often contain sulfur compounds, and their combustion generates sulfur dioxide. Further oxidation of SO_2, usually in the presence of a catalyst such as NO_2, forms H_2SO_4, and thus acid rain. This is one of the causes for concern over the environmental impact of the use of these fuels as power sources.

- Nitrogen oxides (NO_x) - Nitrogen oxides, particularly nitrogen dioxide, are expelled from high temperature combustion, and are also produced during thunderstorms by electric discharge. They can be seen as a brown haze dome above or a plume downwind of cities. Nitrogen dioxide is a chemical compound with the formula NO_2. It is one of several nitrogen oxides. One of the most prominent air pollutants, this reddish-brown toxic gas has a characteristic sharp, biting odor.

- Carbon monoxide (CO) - CO is a colorless, odorless, toxic yet non-irritating gas. It is a product of incomplete combustion of fuel such as natural gas, coal or wood. Vehicular exhaust is a major source of carbon monoxide.

- Volatile organic compounds (VOC) - VOCs are a well-known outdoor air pollutant. They are categorized as either methane (CH_4) or non-methane (NMVOCs). Methane is an extremely efficient greenhouse gas which contributes to enhanced global warming. Other hydrocarbon VOCs are also significant greenhouse gases because of their role in creating ozone and prolonging the life of methane in the atmosphere. This effect varies depending on local air quality. The aromatic NMVOCs benzene, toluene and xylene are suspected carcinogens and may lead to leukemia with prolonged exposure. 1,3-butadiene is another dangerous compound often associated with industrial use.

- Particulates, alternatively referred to as particulate matter (PM), atmospheric particulate matter, or fine particles, are tiny particles of solid or liquid suspended in a gas. In contrast, aerosol refers to combined particles and gas. Some particulates occur naturally, originating from volcanoes, dust storms, forest and grass-land fires, living vegetation, and sea spray. Human activities, such as the burning of fossil fuels in vehicles, power plants and various industrial processes also generate significant amounts of aerosols. Averaged worldwide, anthropogenic aerosols—those made by human activities—currently account for approximately 10 percent of our atmosphere. Increased levels of fine particles in the air are linked to health hazards such as heart disease, altered lung function and lung cancer.

- Persistent free radicals connected to airborne fine particles are linked to cardio-pulmonary disease.

- Toxic metals, such as lead and mercury, especially their compounds.

- Chlorofluorocarbons (CFCs) - harmful to the ozone layer; emitted from products are currently banned from use. These are gases which are released from air conditioners, refrigerators, aerosol sprays, etc. On release into the air, CFCs rise to the stratosphere. Here they come in contact with other gases and damage the ozone layer. This allows harmful ultraviolet rays to reach the earth's surface. This can lead to skin cancer, eye disease and can even cause damage to plants.

- Ammonia (NH_3) - emitted from agricultural processes. Ammonia is a compound with the formula NH_3. It is normally encountered as a gas with a characteristic pungent odor. Ammonia contributes significantly to the nutritional needs of terrestrial organisms by serving as a precursor to foodstuffs and fertilizers. Ammonia, either directly or indirectly, is also a building block for the synthesis of many pharmaceuticals. Although in wide use, ammonia is both caustic and hazardous. In the atmosphere, ammonia reacts with oxides of nitrogen and sulfur to form secondary particles.

- Odours — such as from garbage, sewage, and industrial processes

- Radioactive pollutants - produced by nuclear explosions, nuclear events, war explosives, and natural processes such as the radioactive decay of radon.

Secondary pollutants include:

- Particulates created from gaseous primary pollutants and compounds in photo-chemical smog. Smog is a kind of air pollution. Classic smog results from large amounts of coal burning in an area caused by a mixture of smoke and sulfur dioxide. Modern smog does not usually come from coal but from vehicular and industrial emissions that are acted on in the atmosphere by ultraviolet light from the sun to form secondary pollutants that also combine with the primary emissions to form photochemical smog.

- Ground level ozone (O_3) formed from NO_x and VOCs. Ozone (O_3) is a key constituent of the troposphere. It is also an important constituent of certain regions of the stratosphere commonly known as the Ozone layer. Photochemical and chemical reactions involving it drive many of the chemical processes that occur in the atmosphere by day and by night. At abnormally high concentrations brought about by human activities (largely the combustion of fossil fuel), it is a pollutant, and a constituent of smog.

- Peroxyacetyl nitrate ($C_2H_3NO_5$) - similarly formed from NO_x and VOCs.

Minor air pollutants include:

- A large number of minor hazardous air pollutants. Some of these are regulated in USA under the Clean Air Act and in Europe under the Air Framework Directive

- A variety of persistent organic pollutants, which can attach to particulates

Persistent organic pollutants (POPs) are organic compounds that are resistant to environmental degradation through chemical, biological, and photolytic processes. Because of this, they have been observed to persist in the environment, to be capable of long-range transport, bioaccumulate in human and animal tissue, biomagnify in food chains, and to have potentially significant impacts on human health and the environment.

Sources

There are various locations, activities or factors which are responsible for releasing pollutants into the atmosphere. These sources can be classified into two major categories.

Anthropogenic (Man-made) Sources

Controlled burning of a field outside of Statesboro,
Georgia in preparation for spring planting.

These are mostly related to the burning of multiple types of fuel.

- Stationary sources include smoke stacks of power plants, manufacturing facilities (factories) and waste incinerators, as well as furnaces and other types of fuel-burning heating devices. In developing and poor countries, traditional bio-mass burning is the major source of air pollutants; traditional biomass includes wood, crop waste and dung.

- Mobile sources include motor vehicles, marine vessels, and aircraft.

- Controlled burn practices in agriculture and forest management. Controlled or prescribed burning is a technique sometimes used in forest management, farming, prairie restoration or greenhouse gas abatement. Fire is a natural part of both forest and grassland ecology and controlled fire can be a tool for foresters. Controlled burning stimulates the germination of some desirable forest trees, thus renewing the forest.

- Fumes from paint, hair spray, varnish, aerosol sprays and other solvents

- Waste deposition in landfills, which generate methane. Methane is highly flammable and may form explosive mixtures with air. Methane is also an asphyxiant and may displace oxygen in an enclosed space. Asphyxia or suffocation may result if the oxygen concentration is reduced to below 19.5% by displacement.

- Military resources, such as nuclear weapons, toxic gases, germ warfare and rocketry

Natural Sources

Dust storm approaching Stratford, Texas

- Dust from natural sources, usually large areas of land with little or no vegetation

- Methane, emitted by the digestion of food by animals, for example cattle

- Radon gas from radioactive decay within the Earth's crust. Radon is a colorless, odorless, naturally occurring, radioactive noble gas that is formed from the decay of radium. It is considered to be a health hazard. Radon gas from natural sources can accumulate in buildings, especially in confined areas such as the basement and it is the second most frequent cause of lung cancer, after cigarette smoking.

- Smoke and carbon monoxide from wildfires

- Vegetation, in some regions, emits environmentally significant amounts of Volatile organic compounds (VOCs) on warmer days. These VOCs react with primary anthropogenic pollutants—specifically, NO_x, SO_2, and anthropogenic organic carbon compounds — to produce a seasonal haze of secondary pollutants. Black gum, poplar, oak and willow are some examples of vegetation that can produce abundant VOCs. The VOC production from these species result in ozone levels up to eight times higher than the low-impact tree species.

- Volcanic activity, which produces sulfur, chlorine, and ash particulates

Emission Factors

Beijing air on a 2005-day after rain (left) and a smoggy day (right)

Air pollutant emission factors are reported representative values that attempt to relate the quantity of a pollutant released to the ambient air with an activity associated with the release of that pollutant. These factors are usually expressed as the weight of pollutant divided by a unit weight, volume, distance, or duration of the activity emitting the pollutant (e.g., kilograms of particulate emitted per tonne of coal burned). Such factors facilitate estimation of emissions from various sources of air pollution. In most cases, these factors are simply averages of all available data of acceptable quality, and are generally assumed to be representative of long-term averages.

There are 12 compounds in the list of persistent organic pollutants. Dioxins and furans are two of them and intentionally created by combustion of organics, like open burning of plastics. These compounds are also endocrine disruptors and can mutate the human genes.

The United States Environmental Protection Agency has published a compilation of air pollutant emission factors for a wide range of industrial sources. The United Kingdom, Australia, Canada and many other countries have published similar compilations, as well as the European Environment Agency.

Exposure

Air pollution risk is a function of the hazard of the pollutant and the exposure to that pollutant. Air pollution exposure can be expressed for an individual, for certain groups (e.g. neighborhoods or children living in a country), or for entire populations. For example, one may want to calculate the exposure to a hazardous air pollutant for a geographic area, which includes the various microenvironments and age groups. This can be calculated as an inhalation exposure. This would account for daily exposure in various settings (e.g. different indoor micro-environments and outdoor locations). The exposure needs to include different age and other demographic groups, especially infants, children, pregnant women and other sensitive subpopulations. The exposure to an air pollutant must integrate the concentrations of the air pollutant with respect to the time spent in each setting and the respective inhalation rates for each subgroup for each specific time that the subgroup is in the setting and engaged in particular activities (playing, cooking, reading, working, etc.). For example, a small child's inhalation rate will be less than that of an adult. A child engaged in vigorous exercise will have a higher respiration rate than the same child in a sedentary activity. The daily exposure, then, needs to reflect the time spent in each micro-environmental setting and the type of activities in these settings. The air pollutant concentration in each microactivity/microenvironmental setting is summed to indicate the exposure.

Indoor Air Quality (IAQ)

Air quality monitoring, New Delhi, India.

A lack of ventilation indoors concentrates air pollution where people often spend the majority of their time. Radon (Rn) gas, a carcinogen, is exuded from the Earth in certain locations and trapped inside houses. Building materials including carpeting and plywood emit formaldehyde (H_2CO) gas. Paint and solvents give off volatile organic compounds (VOCs) as they dry. Lead paint can degenerate into dust and be inhaled. Intentional air pollution is introduced with the use of air fresheners, incense, and other scented items. Controlled wood fires in stoves and fireplaces can add significant amounts of smoke particulates into the air, inside and out. Indoor pollution fatalities may be caused by using pesticides and other chemical sprays indoors without proper ventilation.

Carbon monoxide poisoning and fatalities are often caused by faulty vents and chimneys, or by the burning of charcoal indoors or in a confined space, such as a tent. Chronic carbon monoxide poisoning can result even from poorly-adjusted pilot lights. Traps are built into all domestic plumbing to keep sewer gas and hydrogen sulfide, out of interiors. Clothing emits tetrachloroethylene, or other dry cleaning fluids, for days after dry cleaning.

Though its use has now been banned in many countries, the extensive use of asbestos in industrial and domestic environments in the past has left a potentially very dangerous material in many localities. Asbestosis is a chronic inflammatory medical condition affecting the tissue of the lungs. It occurs after long-term, heavy exposure to asbestos from asbestos-containing materials in structures. Sufferers have severe dyspnea (shortness of breath) and are at an increased risk regarding several different types of lung cancer. As clear explanations are not always stressed in non-technical literature, care should be taken to distinguish between several forms of relevant diseases. According to the World Health Organisation (WHO), these may defined as; asbestosis, *lung cancer*, and *Peritoneal Mesothelioma* (generally a very rare form of cancer, when more widespread it is almost always associated with prolonged exposure to asbestos).

Biological sources of air pollution are also found indoors, as gases and airborne particulates. Pets produce dander, people produce dust from minute skin flakes and decomposed hair, dust mites in bedding, carpeting and furniture produce enzymes and micrometre-sized fecal droppings, inhabitants emit methane, mold forms on walls and generates mycotoxins and spores, air conditioning systems can incubate Legionnaires' disease and mold, and houseplants, soil and surrounding gardens can produce pollen, dust, and mold. Indoors, the lack of air circulation allows these airborne pollutants to accumulate more than they would otherwise occur in nature.

Health Effects

Air pollution is a significant risk factor for a number of pollution-related diseases and health conditions including respiratory infections, heart disease, COPD, stroke and lung cancer. The health effects caused by air pollution may include difficulty in breathing, wheezing, coughing, asthma and worsening of existing respiratory and cardiac conditions. These effects can result in increased medication use, increased doctor or emergency room visits, more hospital admissions and premature death. The human health effects of poor air quality are far reaching, but principally affect the body's respiratory system and the cardiovascular system. Individual reactions to air pollutants depend on the type of pollutant a person is exposed to, the degree of exposure, and the individual's health status and genetics. The most common sources of air pollution include particulates, ozone, nitrogen dioxide, and sulphur dioxide. Children aged less than five years that live in developing countries are the most vulnerable population in terms of total deaths attributable to indoor and outdoor air pollution.

Mortality

The World Health Organization estimated in 2014 that every year air pollution causes the premature death of some 7 million people worldwide. India has the highest death rate due to air pollution. India also has more deaths from asthma than any other nation according to the World Health Organization. In December 2013 air pollution was estimated to kill 500,000 people in China each year. There is a positive correlation between pneumonia-related deaths and air pollution from motor vehicle emissions.

Annual premature European deaths caused by air pollution are estimated at 430,000. An important cause of these deaths is nitrogen dioxide and other nitrogen oxides (NOx) emitted by road vehicles. Across the European Union, air pollution is estimated to reduce life expectancy by almost nine months. Causes of deaths include strokes, heart disease, COPD, lung cancer, and lung infections.

Urban outdoor air pollution is estimated to cause 1.3 million deaths worldwide per year. Children are particularly at risk due to the immaturity of their respiratory organ systems.

The US EPA estimates that a proposed set of changes in diesel engine technology (*Tier 2*) could result in 12,000 fewer *premature mortalities*, 15,000 fewer heart attacks, 6,000 fewer emergency room visits by children with asthma, and 8,900 fewer respiratory-related hospital admissions each year in the United States.

The US EPA has estimated that limiting ground-level ozone concentration to 65 parts per billion, would avert 1,700 to 5,100 premature deaths nationwide in 2020 compared with the 75-ppb standard. The agency projected the more protective standard would also prevent an additional 26,000 cases of aggravated asthma, and more than a million cases of missed work or school. Following this assessment, the EPA acted to protect public health by lowering the National Ambient Air Quality Standards (NAAQS) for ground-level ozone to 70 parts per billion (ppb).

A new economic study of the health impacts and associated costs of air pollution in the Los Angeles Basin and San Joaquin Valley of Southern California shows that more than 3,800 people die prematurely (approximately 14 years earlier than normal) each year because air pollution levels violate federal standards. The number of annual premature deaths is considerably higher than the fatalities related to auto collisions in the same area, which average fewer than 2,000 per year.

Diesel exhaust (DE) is a major contributor to combustion-derived particulate matter air pollution. In several human experimental studies, using a well-validated exposure chamber setup, DE has been linked to acute vascular dysfunction and increased thrombus formation.

The mechanisms linking air pollution to increased cardiovascular mortality are uncertain, but probably include pulmonary and systemic inflammation.

Cardiovascular Disease

A 2007 review of evidence found ambient air pollution exposure is a risk factor correlating with increased total mortality from cardiovascular events (range: 12% to 14% per 10 microg/m³ increase).

Air pollution is also emerging as a risk factor for stroke, particularly in developing countries where pollutant levels are highest. A 2007 study found that in women, air pollution is not associated with hemorrhagic but with ischemic stroke. Air pollution was also found to be associated with increased incidence and mortality from coronary stroke in a cohort study in 2011. Associations are believed to be causal and effects may be mediated by vasoconstriction, low-grade inflammation and atherosclerosis Other mechanisms such as autonomic nervous system imbalance have also been suggested.

Lung Disease

Chronic obstructive pulmonary disease (COPD) includes diseases such as chronic bronchitis and emphysema.

Research has demonstrated increased risk of developing asthma and COPD from increased exposure to traffic-related air pollution. Additionally, air pollution has been associated with increased hospitalization and mortality from asthma and COPD.

A study conducted in 1960-1961 in the wake of the Great Smog of 1952 compared 293 London residents with 477 residents of Gloucester, Peterborough, and Norwich, three towns with low reported death rates from chronic bronchitis. All subjects were male postal truck drivers aged 40 to 59. Compared to the subjects from the outlying towns, the London subjects exhibited more severe respiratory symptoms (including cough, phlegm, and dyspnea), reduced lung function (FEV_1 and peak flow rate), and increased sputum production and purulence. The differences were more pronounced for subjects aged 50 to 59. The study controlled for age and smoking habits, so concluded that air pollution was the most likely cause of the observed differences.

It is believed that much like cystic fibrosis, by living in a more urban environment serious health hazards become more apparent. Studies have shown that in urban areas patients suffer mucus hypersecretion, lower levels of lung function, and more self-diagnosis of chronic bronchitis and emphysema.

Cancer

A review of evidence regarding whether ambient air pollution exposure is a risk factor for cancer in 2007 found solid data to conclude that long-term exposure to PM2.5 (fine particulates) increases the overall risk of non-accidental mortality by 6% per a 10 microg/m³ increase. Exposure to PM2.5 was also associated with an increased risk of mortality from lung cancer (range: 15% to 21% per 10 microg/m³ increase) and total cardiovascular mortality (range: 12% to 14% per a 10 microg/m³ increase). The review further noted that living close to busy traffic appears to be associated with elevated risks of these three outcomes --- increase in lung cancer deaths, cardiovascular deaths, and overall non-accidental deaths. The reviewers also found suggestive evidence that exposure to PM2.5 is positively associated with mortality from coronary heart diseases and exposure to SO_2 increases mortality from lung cancer, but the data was insufficient to provide solid conclusions. Another investigation showed that higher activity level increases deposition fraction of aerosol particles in human lung and recommended avoiding heavy activities like running in outdoor space at polluted areas.

Cancer mainly the result of environmental factors.

In 2011, a large Danish epidemiological study found an increased risk of lung cancer for patients who lived in areas with high nitrogen oxide concentrations. In this study, the association was higher for non-smokers than smokers. An additional Danish study, also in 2011, likewise noted evidence of possible associations between air pollution and other forms of cancer, including cervical cancer and brain cancer.

In December 2015, medical scientists reported that cancer is overwhelmingly a result of environmental factors, and not largely down to bad luck. Maintaining a healthy weight, eating a healthy diet, minimizing alcohol and eliminating smoking reduces the risk of developing the disease, according to the researchers.

Children

In the United States, despite the passage of the Clean Air Act in 1970, in 2002 at least 146 million Americans were living in non-attainment areas—regions in which the concentration of certain air pollutants exceeded federal standards. These dangerous pollutants are known as the criteria pollutants, and include ozone, particulate matter, sulfur dioxide, nitrogen dioxide, carbon monoxide, and lead. Protective measures to ensure children's health are being taken in cities such as New Delhi, India where buses now use compressed natural gas to help eliminate the "pea-soup" smog. A recent study in Europe has found that exposure to ultrafine particles can increase blood pressure in children.

Infants

Ambient levels of air pollution have been associated with preterm birth and low birth weight. A 2014 WHO worldwide survey on maternal and perinatal health found a statistically significant association between low birth weights (LBW) and increased levels of exposure to PM2.5. Women in regions with greater than average PM2.5 levels had statistically significant higher odds of pregnancy resulting in a low-birth weight infant even when adjusted for country-related variables. The effect is thought to be from stimulating inflammation and increasing oxidative stress.

A study by the University of York found that in 2010 exposure to PM2.5 was strongly associated with 18% of preterm births globally, which was approximately 2.7 million premature births. The countries with the highest air pollution associated preterm births were in South and East Asia, the Middle East, North Africa, and West sub-Saharan Africa.

The source of PM 2.5 differs greatly by region. In South and East Asia, pregnant women are frequently exposed to indoor air pollution because of the wood and other biomass fuels used for cooking which are responsible for more than 80% of regional pollution. In the Middle East, North Africa and West sub-Saharan Africa, fine PM comes from natural sources, such as dust storms. The United States had an estimated 50,000 preterm births associated with exposure to PM2.5 in 2010.

A study performed by Wang, et al. between the years of 1988 and 1991 has found a correlation between Sulfur Dioxide (SO_2) and total suspended particulates (TSP) and preterm births and low birth weights in Beijing. A group of 74,671 pregnant women, in four separate regions of Beijing, were monitored from early pregnancy to delivery along with daily air pollution levels of Sulfur Dioxide and TSP (along with other particulates). The estimated reduction in birth weight was 7.3 g for every 100 $\mu g/m^3$ increase in SO_2 and 6.9g for each 100 $\mu g/m^3$ increase in TSP. These associations were statistically significant in both summer and winter, although, summer was greater. The proportion of low birth weight attributable to air pollution, was 13%. This is the largest attributable risk ever reported for the known risk factors of low birth weight. Coal stoves, which are in 97% of homes, are a major source of air pollution in this area.

Brauer et al. studied the relationship between air pollution and proximity to a highway with pregnancy outcomes in a Vancouver cohort of pregnant woman using addresses to estimate exposure during pregnancy. Exposure to NO, NO_2, CO PM10 and PM2.5 were associated with infants born small for gestational age (SGA). Women living <50meters away from an expressway or highway were 26% more likely to give birth to a SGA infant.

"Clean" areas

Even in the areas with relatively low levels of air pollution, public health effects can be significant and costly, since a large number of people breathe in such pollutants. A 2005 scientific study for the British Columbia Lung Association showed that a small improvement in air quality (1% reduction of ambient PM2.5 and ozone concentrations) would produce $29 million in annual savings in the Metro Vancouver region in 2010. This finding is based on health valuation of lethal (death) and sub-lethal (illness) affects.

Central Nervous System

Data is accumulating that air pollution exposure also affects the central nervous system.

In a June 2014 study conducted by researchers at the University of Rochester Medical Center, published in the journal Environmental Health Perspectives, it was discovered that early exposure to air pollution causes the same damaging changes in the brain as autism and schizophrenia. The study also shows that air pollution also affected short-term memory, learning ability, and impulsivity. Lead researcher Professor Deborah Cory-Slechta said that "When we looked closely at the ventricles, we could see that the white matter that normally surrounds them hadn't fully developed. It appears that inflammation had damaged those brain cells and prevented that region of the brain from developing, and the ventricles simply expanded to fill the space. Our findings add to the growing body of evidence that air pollution may play a role in autism, as well as in other neurodevelopmental disorders." Air pollution has a more significant negative effect on males than on females.

In 2015, experimental studies reported the detection of significant episodic (situational) cognitive impairment from impurities in indoor air breathed by test subjects who were not informed about changes in the air quality. Researchers at the Harvard University and SUNY Upstate Medical University and Syracuse University measured the cognitive performance of 24 participants in three different controlled laboratory atmospheres that simulated those found in "conventional" and "green" buildings, as well as green buildings with enhanced ventilation. Performance was evaluated objectively using the widely used Strategic Management Simulation software simulation tool, which is a well-validated assessment test for executive decision-making in an unconstrained situation allowing initiative and improvisation. Significant deficits were observed in the performance scores achieved in increasing concentrations of either volatile organic compounds (VOCs) or carbon dioxide, while keeping other factors constant. The highest impurity levels reached are not uncommon in some classroom or office environments.

Agricultural Effects

In India in 2014, it was reported that air pollution by black carbon and ground level ozone had cut crop yields in the most affected areas by almost half in 2010 when compared to 1980 levels.

Economic Effects

Air pollution costs the world economy $5 trillion per year as a result of productivity losses and degraded quality of life, according to a joint study by the World Bank and the Institute for Health Metrics and Evaluation (IHME) at the University of Washington These productivity losses are caused by deaths due to diseases caused by air pollution. One out of ten deaths in 2013 was caused by diseases associated with air pollution and the problem is getting worse. The problem is even more acute in the developing world. "Children under age 5 in lower-income countries are more than 60 times as likely to die from exposure to air pollution as children in high-income countries." The report states that additional economic losses caused by air pollution, including health costs and the adverse effect on agricultural and other productivity were not calculated in the report, and thus the actual costs to the world economy are far higher than $5 trillion.

Historical Disasters

The world's worst short-term civilian pollution crisis was the 1984 Bhopal Disaster in India. Leaked industrial vapours from the Union Carbide factory, belonging to Union Carbide, Inc., U.S.A. (later bought by Dow Chemical Company), killed at least 3787 people and injured from 150,000 to 600,000. The United Kingdom suffered its worst air pollution event when the December 4 Great Smog of 1952 formed over London. In six days more than 4,000 died and more recent estimates put the figure at nearer 12,000. An accidental leak of anthrax spores from a biological warfare laboratory in the former USSR in 1979 near Sverdlovsk is believed to have caused at least 64 deaths. The worst single incident of air pollution to occur in the US occurred in Donora, Pennsylvania in late October, 1948, when 20 people died and over 7,000 were injured.

Alternatives to Pollution

There are now practical alternatives to the principal causes of air pollution:

- Areas downwind (over 20 miles) of major airports more than double *total particulate emissions in air*, even when factoring in areas with frequent ship calls, and heavy freeway and city traffic like Los Angeles. Aviation biofuel mixed in with jetfuel at a 50/50 ratio can reduce jet derived cruise altitude particulate emissions by 50-70%, according to a NASA led 2017 study (however, this should imply ground level benefits to urban air pollution as well).

- Ship propulsion and idling can be switched to much cleaner fuels like natural gas. (Ideally a renewable source but not practical yet)

- Combustion of fossil fuels for space heating can be replaced by using ground source heat pumps and seasonal thermal energy storage.

- Electric power generation from burning fossil fuels can be replaced by power generation from nuclear and renewables. For poor nations, heating and home stoves that contribute much to regional air pollution can be replaced by a much cleaner fossil fuel like natural gas, or ideally, renewables.

- Motor vehicles driven by fossil fuels, a key factor in urban air pollution, can be replaced by electric vehicles. Though lithium supply and cost is a limitation, there are alternatives. Herding more people into clean public transit such as electric trains can also help. Nevertheless, even in emission-free electric vehicles, rubber tires produce significant amounts of air pollution themselves, ranking as 13th worst pollutant in Los Angeles.

- Biodigesters can be utilized in poor nations where slash and burn is prevalent, turning a useless commodity into a source of income. The plants can be gathered and sold to a central authority that will break it down in a large modern biodigester, producing much needed energy to use.

- Induced humidity and ventilation both can greatly dampen air pollution in enclosed spaces, which was found to be relatively high inside subway lines due to braking and friction and relatively less ironically inside transit buses than lower sitting passenger automobiles or subways.

Reduction Efforts

There are various air pollution control technologies and strategies available to reduce air pollution. At its most basic level, land-use planning is likely to involve zoning and transport infrastructure planning. In most developed countries, land-use planning is an important part of social policy, ensuring that land is used efficiently for the benefit of the wider economy and population, as well as to protect the environment.

Because a large share of air pollution is caused by combustion of fossil fuels such as coal and oil, the reduction of these fuels can reduce air pollution drastically. Most effective is the switch to clean power sources such as wind power, solar power, hydro power which don't cause air pollution. Efforts to reduce pollution from mobile sources includes primary regulation (many developing countries have permissive regulations), expanding regulation to new sources (such as cruise and transport ships, farm equipment, and small gas-powered equipment such as string trimmers, chainsaws, and snowmobiles), increased fuel efficiency (such as through the use of hybrid vehicles), conversion to cleaner fuels or conversion to electric vehicles.

Titanium dioxide has been researched for its ability to reduce air pollution. Ultraviolet light will release free electrons from material, thereby creating free radicals, which break up VOCs and NOx gases. One form is superhydrophilic.

In 2014, Prof. Tony Ryan and Prof. Simon Armitage of University of Sheffield prepared a 10 meter by 20 meter-sized poster coated with microscopic, pollution-eating nanoparticles of titanium dioxide. Placed on a building, this giant poster can absorb the toxic emission from around 20 cars each day.

A very effective means to reduce air pollution is the transition to renewable energy. According to a study published in Energy and Environmental Science in 2015 the switch to 100% renewable energy in the United States would eliminate about 62,000 premature mortalities per year and about 42,000 in 2050, if no biomass were used. This would save about $600 billion in health costs a year due to reduced air pollution in 2050, or about 3.6% of the 2014 U.S. gross domestic product.

Control Devices

The following items are commonly used as pollution control devices in industry and transportation. They can either destroy contaminants or remove them from an exhaust stream before it is emitted into the atmosphere.

- Particulate control

    o   Mechanical collectors (dust cyclones, multicyclones)

    o   Electrostatic precipitators An electrostatic precipitator (ESP), or electrostatic air cleaner is a particulate collection device that removes particles from a flowing gas (such as air), using the force of an induced electrostatic charge. Electrostatic precipitators are highly efficient filtration devices that minimally impede the flow of gases through the device, and can easily remove fine particulates such as dust and smoke from the air stream.

    o   Baghouses Designed to handle heavy dust loads, a dust collector consists of a blower, dust filter, a filter-cleaning system, and a dust receptacle or dust removal system (distinguished from air cleaners which utilize disposable filters to remove the dust).

    o   Particulate scrubbers Wet scrubber is a form of pollution control technology. The term describes a variety of devices that use pollutants from a furnace flue gas or from other gas streams. In a wet scrubber, the polluted gas stream is brought into contact with the scrubbing liquid, by spraying it with the liquid, by forcing it through a pool of liquid, or by some other contact method, so as to remove the pollutants.

- Scrubbers

    o   Baffle spray scrubber

    o   Cyclonic spray scrubber

- o Ejector venturi scrubber

- o Mechanically aided scrubber

- o Spray tower

- o Wet scrubber

- NOx control

  - o Low NOx burners

  - o Selective catalytic reduction (SCR)

  - o Selective non-catalytic reduction (SNCR)

  - o NOx scrubbers

  - o Exhaust gas recirculation

  - o Catalytic converter (also for VOC control)

- VOC abatement

  - o Adsorption systems, using activated carbon, such as Fluidized Bed Concentrator

  - o Flares

  - o Thermal oxidizers

  - o Catalytic converters

  - o Biofilters

  - o Absorption (scrubbing)

  - o Cryogenic condensers

  - o Vapor recovery systems

- Acid Gas/SO_2 control

  - o Wet scrubbers

  - o Dry scrubbers

  - o Flue-gas desulfurization

- Mercury control

  - o Sorbent Injection Technology

- o   Electro-Catalytic Oxidation (ECO)

- o   K-Fuel

- • Dioxin and furan control

- • Miscellaneous associated equipment

  - o   Source capturing systems

  - o   Continuous emissions monitoring systems (CEMS)

Regulations

Smog in Cairo

In general, there are two types of air quality standards. The first class of standards (such as the U.S. National Ambient Air Quality Standards and E.U. Air Quality Directive) set maximum atmospheric concentrations for specific pollutants. Environmental agencies enact regulations which are intended to result in attainment of these target levels. The second class (such as the North American Air Quality Index) take the form of a scale with various thresholds, which is used to communicate to the public the relative risk of outdoor activity. The scale may or may not distinguish between different pollutants.

Canada

In Canada, air pollution and associated health risks are measured with the Air Quality Health Index or (AQHI). It is a health protection tool used to make decisions to reduce short-term exposure to air pollution by adjusting activity levels during increased levels of air pollution.

The Air Quality Health Index or "AQHI" is a federal program jointly coordinated by Health Canada and Environment Canada. However, the AQHI program would not be possible without the commitment and support of the provinces, municipalities and NGOs. From air quality monitoring to health risk communication and community engagement, local partners are responsible for the vast majority of work related to AQHI implementation. The AQHI provides a number from 1 to 10+ to indicate the level of health risk associated with local air quality. Occasionally, when the amount of air pollution is abnormally high, the number may exceed 10. The AQHI provides a local air quality current value as well as a local air quality maximums forecast for today, tonight and tomorrow and provides associated health advice.

Risk: Low (1-3)   Moderate (4-6)   High (7-10)   Very high (above 10)

As it is now known that even low levels of air pollution can trigger discomfort for the sensitive population, the index has been developed as a continuum: The higher the number, the greater the health risk and need to take precautions. The index describes the level of health risk associated with this number as 'low', 'moderate', 'high' or 'very high', and suggests steps that can be taken to reduce exposure.

Health Risk	Air Quality Health Index	Health Messages	
		At Risk population	General Population
Low	1-3	Enjoy your usual outdoor activities.	Ideal air quality for outdoor activities
Moderate	4-6	Consider reducing or rescheduling strenuous activities outdoors if you are experiencing symptoms.	No need to modify your usual outdoor activities unless you experience symptoms such as coughing and throat irritation.
High	7-10	Reduce or reschedule strenuous activities outdoors. Children and the elderly should also take it easy.	Consider reducing or rescheduling strenuous activities outdoors if you experience symptoms such as coughing and throat irritation.
Very high	Above 10	Avoid strenuous activities outdoors. Children and the elderly should also avoid outdoor physical exertion and should stay indoors.	Reduce or reschedule strenuous activities outdoors, especially if you experience symptoms such as coughing and throat irritation.

The measurement is based on the observed relationship of Nitrogen Dioxide (NO_2), ground-level Ozone (O_3) and particulates ($PM_{2.5}$) with mortality, from an analysis of several Canadian cities. Significantly, all three of these pollutants can pose health risks, even at low levels of exposure, especially among those with pre-existing health problems.

When developing the AQHI, Health Canada's original analysis of health effects included five major air pollutants: particulates, ozone, and nitrogen dioxide (NO_2), as well as sulfur dioxide (SO_2), and carbon monoxide (CO). The latter two pollutants provided little information in predicting health effects and were removed from the AQHI formulation.

The AQHI does not measure the effects of odour, pollen, dust, heat or humidity.

Germany

TA Luft is the German air quality regulation.

Hotspots

Air pollution hotspots are areas where air pollution emissions expose individuals to increased negative health effects. They are particularly common in highly populated, urban areas, where there may be a combination of stationary sources (e.g. industrial facilities) and mobile sources (e.g. cars and trucks) of pollution. Emissions from these sources can cause respiratory disease, childhood asthma, cancer, and other health problems. Fine particulate matter such as diesel soot, which contributes to more than 3.2 million premature deaths around the world each year, is a significant problem. It is very small and can lodge itself within the lungs and enter the bloodstream. Diesel soot is concentrated in densely populated areas, and one in six people in the U.S. live near a diesel pollution hot spot.

While air pollution hotspots affect a variety of populations, some groups are more likely to be located in hotspots. Previous studies have shown disparities in exposure to pollution by race and/or income. Hazardous land uses (toxic storage and disposal facilities, manufacturing facilities, major roadways) tend to be located where property values and income levels are low. Low socioeconomic status can be a proxy for other kinds of social vulnerability, including race, a lack of ability to influence regulation and a lack of ability to move to neighborhoods with less environmental pollution. These communities bear a disproportionate burden of environmental pollution and are more likely to face health risks such as cancer or asthma.

Studies show that patterns in race and income disparities not only indicate a higher exposure to pollution but also higher risk of adverse health outcomes. Communities characterized by low socioeconomic status and racial minorities can be more vulnerable to cumulative adverse health impacts resulting from elevated exposure to pollutants than more privileged communities. Blacks and Latinos generally face more pollution than whites and Asians, and low-income communities bear a higher burden of risk than affluent ones. Racial discrepancies are particularly distinct in suburban areas of the US South and metropolitan areas of the US West. Residents in public housing, who are generally low-income and cannot move to healthier neighborhoods, are highly affected by nearby refineries and chemical plants.

Cities

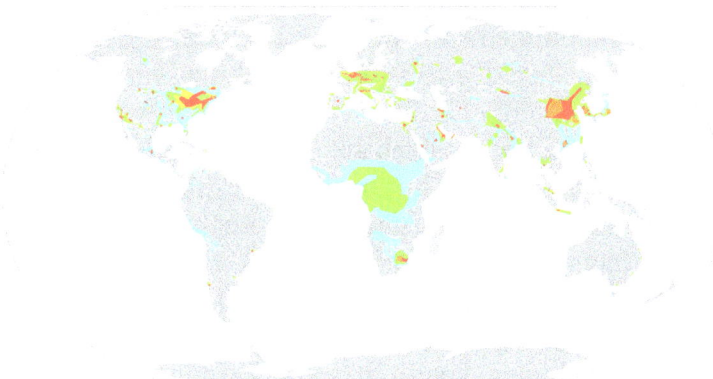

Nitrogen dioxide concentrations as measured from satellite 2002-2004

Air pollution is usually concentrated in densely populated metropolitan areas, especially in developing countries where environmental regulations are relatively lax or nonexistent. However, even populated areas in developed countries attain unhealthy levels of pollution, with Los Angeles and Rome being two examples. Between 2002 and 2011 the incidence of lung cancer in Beijing near doubled. While smoking remains the leading cause of lung cancer in China, the number of smokers is falling while lung cancer rates are rising. Another project focusing on the effects on pollution in vegetation has been researched by the local university in Sheffield, UK.

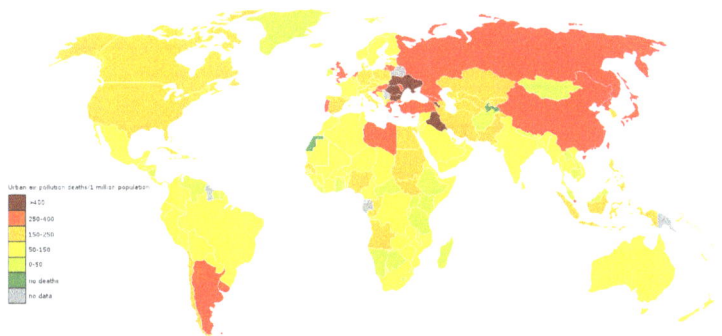

Deaths from air pollution in 2004

National-scale Air Toxics Assessments 1995-2005

The national-scale air toxics assessment(NATA) is an evaluation of air toxics by the U.S. EPA. EPA has furnished four assessments that characterize nationwide chronic cancer risk estimates and noncancer hazards from inhaling air toxics. The lates was from 2005, and made publicly available in early 2011.

"EPA developed the NATA as a state-of-the-science screening tool for State/Local/ Tribal Agencies to prioritize pollutants, emission sources and locations of interest for further study, in order to gain a better understanding of the risks. NATA assessments do not incorporate refined information about emission sources, but rather, use general information about sources to develop estimates of risks which are more likely to overestimate impacts than underestimate them. NATA provides estimates of the risk of cancer and other serious health effects from breathing (inhaling) air toxics in order to inform both national and more localized efforts to identify and prioritize air toxics, emission source types and locations which are of greatest potential concern in terms of contributing to population risk. This in turn helps air pollution experts focus limited analytical resources on areas and or populations where the potential for health risks are highest. Assessments include estimates of cancer and non-cancer health effects based on chronic exposure from outdoor sources, including assessments of non-cancer health effects for Diesel Particulate Matter. Assessments provide a snapshot of the outdoor air quality and the risks to human health that would result if air toxic emissions levels remained unchanged."

Most polluted cities by PM	
Particulate matter, µg/m³ (2004)	City
168	Cairo, Egypt
150	Delhi, India
128	Kolkata, India (Calcutta)
125	Tianjin, China
123	Chongqing, China
109	Kanpur, India
109	Lucknow, India
104	Jakarta, Indonesia
101	Shenyang, China

Governing Urban Air Pollution

In Europe, Council Directive 96/62/EC on ambient air quality assessment and management provides a common strategy against which member states can "set objectives for ambient air quality in order to avoid, prevent or reduce harmful effects on human health and the environment... and improve air quality where it is unsatisfactory".

On 25 July 2008 in the case Dieter Janecek v Freistaat Bayern CURIA, the European Court of Justice ruled that under this directive citizens have the right to require national authorities to implement a short term action plan that aims to maintain or achieve compliance to air quality limit values.

This important case law appears to confirm the role of the EC as centralised regulator to European nation-states as regards air pollution control. It places a supranational legal obligation on the UK to protect its citizens from dangerous levels of air pollution, furthermore superseding national interests with those of the citizen.

In 2010, the European Commission (EC) threatened the UK with legal action against the successive breaching of PM10 limit values. The UK government has identified that if fines are imposed, they could cost the nation upwards of £300 million per year.

In March 2011, the Greater London Built-up Area remains the only UK region in breach of the EC's limit values, and has been given 3 months to implement an emergency action plan aimed at meeting the EU Air Quality Directive. The City of London has dangerous levels of PM10 concentrations, estimated to cause 3000 deaths per year within the city. As well as the threat of EU fines, in 2010 it was threatened with legal action for scrapping the western congestion charge zone, which is claimed to have led to an increase in air pollution levels.

In response to these charges, Boris Johnson, Mayor of London, has criticised the current need for European cities to communicate with Europe through their nation state's central government, arguing that in future "A great city like London" should be permitted to bypass its government and deal directly with the European Commission regarding its air quality action plan.

This can be interpreted as recognition that cities can transcend the traditional national government organisational hierarchy and develop solutions to air pollution using global governance networks, for example through transnational relations. Transnational relations include but are not exclusive to national governments and intergovernmental organisations, allowing sub-national actors including cities and regions to partake in air pollution control as independent actors.

Particularly promising at present are global city partnerships. These can be built into networks, for example the C40 Cities Climate Leadership Group, of which London is a member. The C40 is a public 'non-state' network of the world's leading cities that aims to curb their greenhouse emissions. The C40 has been identified as 'governance from the middle' and is an alternative to intergovernmental policy. It has the potential to improve urban air quality as participating cities "exchange information, learn from best practices and consequently mitigate carbon dioxide emissions independently from national government decisions". A criticism of the C40 network is that its exclusive nature limits influence to participating cities and risks drawing resources away from less powerful city and regional actors.

Air Pollution Control

[A] Mobile Sources

Cleaner/Alternative Fuel

- Vaporization of Gasoline should be reduced.

- Oxygen containing additives reduce air requirement e.g., ethanol, MTBE (Hazardous).

- Methanol: (Less photochemically reactive VOC, but emits HCHO (eye irritant), difficult to start in winters: Can be overcome by M85 (85% methanol, 15% gasoline)

- Ethanol: GASOHOL (10% ethanol & 90% Gasoline),

- CNG: Low HC, NOx high, inconvenient refueling, leakage hazard.

- LPG: Propane, NOx high

Three-Way Catalytic Converter

A three-way catalytic converter has three simultaneous tasks:

- Reduction of nitrogen oxides to nitrogen and oxygen

- Oxidation of carbon monoxide to carbon dioxide

- Oxidation of unburnt hydrocarbons (HC) to carbon dioxide and water

[B] Stationary Sources

Pre-combustion Control

- Switching to less sulphur and nitrogen fuel

Combustion Control

- Improving the combustion process

- New burners to reduce NOx

- New Fluidized bed boilers

- Integrated gasification combined cycle

Post-Combustion Control

- Particulate collection devices

- Flue gas desulphurization

Cyclonic Separation

Cyclonic separation is a method of removing particulates from an air, gas or liquid stream, without the use of filters, through vortex separation. When removing particulate matter from liquids, a hydrocyclone is used; while from gas, a gas cyclone is used. Rotational effects and gravity are used to separate mixtures of solids and fluids. The method can also be used to separate fine droplets of liquid from a gaseous stream.

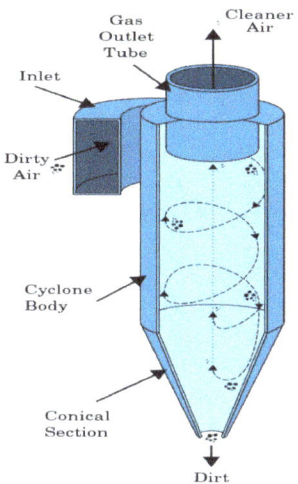

A simple cyclone separator

A high speed rotating (air)flow is established within a cylindrical or conical container called a cyclone. Air flows in a helical pattern, beginning at the top (wide end) of the cyclone and ending at the bottom (narrow) end before exiting the cyclone in a straight stream through the center of the cyclone and out the top. Larger (denser) particles in the rotating stream have too much inertia to follow the tight curve of the stream, and strike the outside wall, then fall to the bottom of the cyclone where they can be removed. In a conical system, as the rotating flow moves towards the narrow end of the cyclone, the rotational radius of the stream is reduced, thus separating smaller and smaller particles. The cyclone geometry, together with flow rate, defines the *cut point* of the cyclone. This is the size of particle that will be removed from the stream with a 50% efficiency. Particles larger than the cut point will be removed with a greater efficiency, and smaller particles with a lower efficiency.

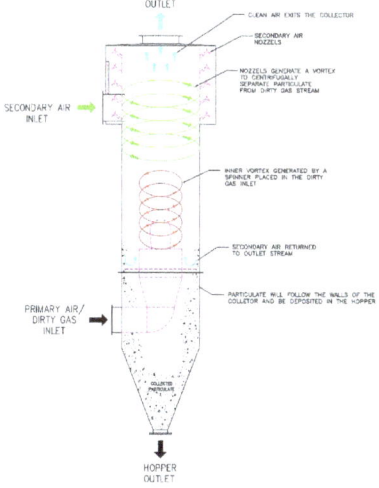

Airflow diagram for Aerodyne cyclone in standard vertical position. Secondary air flow is injected to reduce wall abrasion

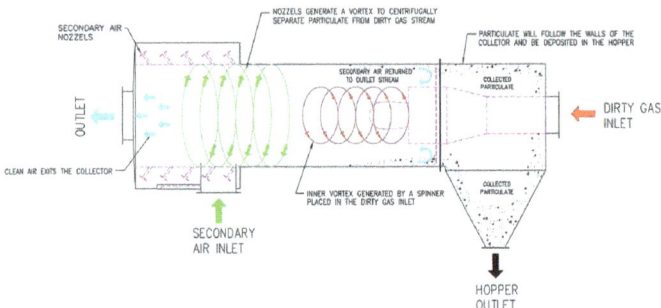

Airflow diagram for Aerodyne cyclone in horizontal position, an alternate design. Secondary air flow is injected to reduce wall abrasion, and to help move collected particulates to hopper for extraction

An alternative cyclone design uses a secondary air flow within the cyclone to keep the collected particles from striking the walls, to protect them from abrasion. The primary air flow containing the particulates enters from the bottom of the cyclone and is forced into spiral rotation by stationary spinner vanes. The secondary air flow enters from the top of the cyclone and moves downward toward the bottom, intercepting the particulate from the primary air. The secondary air flow also allows the collector to optionally be mounted horizontally, because it pushes the particulate toward the collection area, and does not rely solely on gravity to perform this function.

Large scale cyclones are used in sawmills to remove sawdust from extracted air. Cyclones are also used in oil refineries to separate oils and gases, and in the cement industry as components of kiln preheaters. Cyclones are increasingly used in the household, as the core technology in bagless types of portable vacuum cleaners and central vacuum cleaners. Cyclones are also used in industrial and professional kitchen ventilation for separating the grease from the exhaust air in extraction hoods. Smaller cyclones are used to separate airborne particles for analysis. Some are small enough to be worn clipped to clothing, and are used to separate respirable particles for later analysis.

Similar separators are used in the oil refining industry (e.g. for Fluid catalytic cracking) to achieve fast separation of the catalyst particles from the reacting gases and vapors.

James Dyson has become a billionaire from developing and marketing bagless vacuum cleaners based on cyclonic separation of dust, initially inspired by seeing sawdust separator at a sawmill.

Analogous devices for separating particles or solids from liquids are called hydrocyclones or hydroclones. These may be used to separate solid waste from water in wastewater and sewage treatment.

Cyclone Theory

As the cyclone is essentially a two phase particle-fluid system, fluid mechanics and particle transport equations can be used to describe the behaviour of a cyclone. The air in a cyclone is initially introduced tangentially into the cyclone with an inlet velocity V_{in}. Assuming that the particle is spherical, a simple analysis to calculate critical separation particle sizes can be established.

If one considers an isolated particle circling in the upper cylindrical component of the cyclone at a rotational radius of r from the cyclone's central axis, the particle is therefore subjected to drag, centrifugal, and buoyant forces. Given that the fluid velocity is moving in a spiral the gas velocity can be broken into two component velocities: a tangential component, V_t, and an outward radial velocity component V_r. Assuming Stokes' law, the drag force in the outward radial direction that is opposing the outward velocity on any particle in the inlet stream is:

$$F_d = -6\pi r_p \mu V_r.$$

Using ρ_p as the particles density, the centrifugal component in the outward radial direction is:

$$F_c = m \frac{V_t^2}{r}$$

$$= \frac{4}{3} \pi \rho_p r_p^3 \frac{V_t^2}{r}.$$

The buoyant force component is in the inward radial direction. It is in the opposite direction to the particle's centrifugal force because it is on a volume of fluid that is missing compared to the surrounding fluid. Using ρ_f for the density of the fluid, the buoyant force is:

$$F_b = -V_p \rho_f \frac{V_t^2}{r}$$

$$= -\frac{4\pi r_p^3}{3} \frac{V_t^2}{r} \rho_f.$$

In this case, V_p is equal to the volume of the particle (as opposed to the velocity). Determining the outward radial motion of each particle is found by setting Newton's second law of motion equal to the sum of these forces:

$$m \frac{dV_r}{dt} = F_d + F_c + F_b$$

To simplify this, we can assume the particle under consideration has reached "terminal velocity", i.e., that its acceleration $\dfrac{dV_r}{dt}$ is zero. This occurs when the radial velocity has caused enough drag force to counter the centrifugal and buoyancy forces. This simplification changes our equation to:

$$F_d + F_c + F_b = 0$$

Which expands to:

$$-6\pi r_p \mu V_r + \frac{4}{3}\pi r_p^3 \frac{V_t^2}{r}\rho_p - \frac{4}{3}\pi r_p^3 \frac{V_t^2}{r}\rho_f = 0$$

Solving for V_r we have

$$V_r = \frac{2}{9}\frac{r_p^2}{\mu}\frac{V_t^2}{r}(\rho_p - \rho_f)$$

Notice that if the density of the fluid is greater than the density of the particle, the motion is (-), toward the center of rotation and if the particle is denser than the fluid, the motion is (+), away from the center. In most cases, this solution is used as guidance in designing a separator, while actual performance is evaluated and modified empirically.

In non-equilibrium conditions when radial acceleration is not zero, the general equation from above must be solved. Rearranging terms we obtain

$$\frac{dV_r}{dt} + \frac{9}{2}\frac{\mu}{\rho_p r_p^2}V_r - \left(1 - \frac{\rho_f}{\rho_p}\right)\frac{V_t^2}{r} = 0$$

Since V_r is distance per time, this is a 2nd order differential equation of the form $x'' + c_1 x' + c_2 = 0.\cdot$

Experimentally it is found that the velocity component of rotational flow is proportional to r^2, therefore: $V_t \propto r^2$.

This means that the established feed velocity controls the vortex rate inside the cyclone, and the velocity at an arbitrary radius is therefore:

$$U_r = U_{in}\frac{r}{R_{in}}.$$

Subsequently, given a value for V_t, possibly based upon the injection angle, and a cutoff radius, a characteristic particle filtering radius can be estimated, above which particles will be removed from the gas stream.

Alternative Models

The above equations are limited in many regards. For example, the geometry of the separator is not considered, the particles are assumed to achieve a steady state and the effect of the vortex inversion at the base of the cyclone is also ignored, all behaviours which are unlikely to be achieved in a cyclone at real operating conditions.

More complete models exist, as many authors have studied the behaviour of cyclone separators. Numerical modelling using computational fluid dynamics has also been used extensively in the study of cyclonic behaviour. A major limitation of any fluid mechanics model for cyclone separators is the inability to predict the agglomeration of fine particles with larger particles, which has a great impact on cyclone collection efficiency.

Design of Cyclones

Cyclone separators utilizes a centrifugal forces generated by a spinning gas stream to separate the particulate matters from the carrier gas. The centrifugal force on particles in a spinning gas stream is much greater than gravity; therefore cyclones are effective in the removal of much smaller particles than gravitational settling chambers, and require much less space to handle the same gas volumes.

In operation, the particle-laden gas upon entering the cyclone cylinder receives a rotating motion. The vortex so formed develops a centrifugal force, which acts to particle radially towards the wall. The gas spirals downward to the bottom of the cone, and at the bottom the gas flow reveres to form an inner vortex which leaves through the outlet pipe.

Theory

In a cyclone, the inertial separating force is the radial component of the simple centrifugal force and is a function of the tangential velocity. The centrifugal force can be expressed by F_c

$$F_c = \frac{mv_e^2}{r}$$

Where, m=mass of the particle, v_e=tangential velocity of the particle at radius r, and r=radius of rotation. The separation factor S is given by

$$S = \frac{v_e^2}{gr}$$

The separation factor varies from 5 in large, low velocity units to 2500 in small, high pressure units. Higher the separation factor better is the performance of the cyclone.

In the cyclone, the gas, in addition to moving in a circular path, also moves radially inwards between the inlet on the periphery and the exit on the axis. Since the tangential velocities of the particle and the gas are the same, the relative velocity between the gas and particle is simply equal to the radial velocity of the gas. This result in a drag force on the particle towards the centre, and the equilibrium radius of rotation of the particle can be obtained by balancing the radial drag force and the centrifugal force:

$$3\pi\mu_g d_p v_r = \frac{\pi}{6} d_p^3 \left(\rho_p - \rho_g\right) \frac{v_\theta^2}{r}$$

Where, d_p=particle diameter, and v_r=radial velocity of the gas at radius r. Arranging the above equation, for v_r

$$v_r = \frac{d_p^3 \left(\rho_p - \rho_g\right) v_\theta^2}{18\mu_g r}$$

The tangential velocity of the particle in the vortex has been found experimentally to be inversely proportional to the radius of rotation according to equation,

$$v_\theta r^n = \text{constant}$$

Where, n is the exponent and dimensionless. For an ideal gas n=1. The real values observed are between 0.5 to 1, depending upon the radius of the cyclone body and gas temperature. v_θ can be related to the tangential velocity at the inlet to the cyclone $v_{\theta i}$ as

$$v_\theta = v_{\theta i} \left(\frac{D}{2r}\right)^n$$

Where, D=diameter of the cyclone. $v_{\theta i}$ may be taken as the velocity of the gas through the inlet pipe, i.e.,

$$v_{\theta i} = \frac{Q}{A_i}$$

Where, Q=gas volumetric flow rate and A_i=cross-sectional area of the inlet. Therefore,

$$v_\theta = \frac{Q}{A_i}\left(\frac{D}{2r}\right)^n$$

$$v_r = \frac{d_p^3 \left(\rho_p - \rho_g\right)}{18\mu_g r^{(2n+1)}}\left(\frac{Q}{A_i}\right)^2\left(\frac{D}{2r}\right)^{2n}$$

The most satisfactory expression for cyclone performance is still the empirical one.

Lapple correlated collection efficiency in terms of the cut size d_{pe} which is the size of those particle that are collected with 50% efficiency. Particle larger than d_{pe} will have collection efficiency greater than 50% while the smaller particle will be collected with lesser efficiency. The cut size is given by:

$$d_{pe} = \sqrt{\frac{9\mu_g b}{2\pi N_e v_i (\rho_p - \rho_g)}}$$

Where, b=inlet width, v_i=gas inlet velocity and N_e=effective number of turns a gas makes in traversing the cyclone (5 to 10 in most cases).

Pressure drop: The pressure drop may be estimated according to the following equation,

$$\Delta P = \frac{K\rho_g v_i^2 (ab)}{2D_e^2}$$

Where, K=a constant, which averages 13 and ranges from 7.5 to 18.4, ΔP =pressure drop, a, b and D_e=cyclone dimensions, v_i=inlet gas velocity and ρ_g =gas density.

Problem 2.3.1: A conventional cyclone with diameter 0.5 m handles 4.0 m³/s of standard air (μ_g=1.81×10⁻⁵ kg/m-s and ρ_g being negligible w.r.t ρ_p) carrying particles with a density of 2500 kg/m³. For N_e=6, inlet width (b)=0.25 m, inlet height (a)=0.5 m, determine the cut size of particle diameter.

Solution: Given

$$b = 0.25, \qquad D = 0.25 \times 0.5 = 0.1$$
$$a = 0.5, \qquad D = 0.5 \times 0.5 = 0.25$$
$$\rho_\rho = 2500 \,\text{kg/m}^3$$
$$\mu_g = 1.81 \times 10^{-5} \,\text{kg/m-s}$$
$$Q = 4 \,\text{m}^3/\text{s}$$

$$v_i = \frac{Q}{a \times b} = \frac{4}{0.1 \times 0.25} = 160 \,\text{m/s}$$

$$d_{pe} = \sqrt{\frac{9\mu_g b}{2\pi N_e v_i (\rho_p - \rho_g)}}$$

$$d_{pe} = \sqrt{\frac{9 \times 1.81 \times 0.25}{2 \times \pi \times 6 \times 160 \times 2500}} = 5.195 \times 10^{-4} \, m$$

Baghouse

A baghouse (BH, B/H), bag filter (BF) or fabric filter (FF) is an air pollution control device that removes particulates out of air or gas released from commercial processes or combustion for electricity generation. Power plants, steel mills, pharmaceutical producers, food manufacturers, chemical producers and other industrial companies often use baghouses to control emission of air pollutants. Baghouses came into widespread use in the late 1970s after the invention of high-temperature fabrics (for use in the filter media) capable of withstanding temperatures over 350 °F.

Baghouse dust collector for asphalt plants

Unlike electrostatic precipitators, where performance may vary significantly depending on process and electrical conditions, functioning baghouses typically have a particulate collection efficiency of 99% or better, even when particle size is very small.

Operation

Most baghouses use long, cylindrical bags (or tubes) made of woven or felted fabric as a filter medium. (For applications where there is relatively low dust loading and gas temperatures are 250 °F or less, pleated, nonwoven cartridges are sometimes used as filtering media instead of bags.) Dust-laden gas or air enters the baghouse through hoppers (large funnel-shaped containers used for storing and dispensing particulate) and is directed into the baghouse compartment. The gas is drawn through the bags, either on the inside or the outside depending on cleaning method, and a layer of dust accumulates on the filter media surface until air can no longer move through it. When sufficient pressure drop (delta P) occurs, the cleaning process begins. Cleaning can take place while the baghouse is online (filtering) or is offline (in isolation). When the compartment is clean, normal filtering resumes.

Baghouses are very efficient particulate collectors because of the dust cake formed on the surface of the bags. The fabric provides a surface on which dust collects through the following four mechanisms:

- Inertial collection - Dust particles strike the fibers placed perpendicular to the gas-flow direction instead of changing direction with the gas stream.

- Interception - Particles that do not cross the fluid streamlines come in contact with fibers because of the fiber size.

- Brownian movement - Submicrometre particles are diffused, increasing the probability of contact between the particles and collecting surfaces.

- Electrostatic forces - The presence of an electrostatic charge on the particles and the filter can increase dust capture.

A combination of these mechanisms results in formation of the dust cake on the filter, which eventually increases the resistance to gas flow. The filter must be cleaned periodically.

Baghouse types - Cleaning Methods

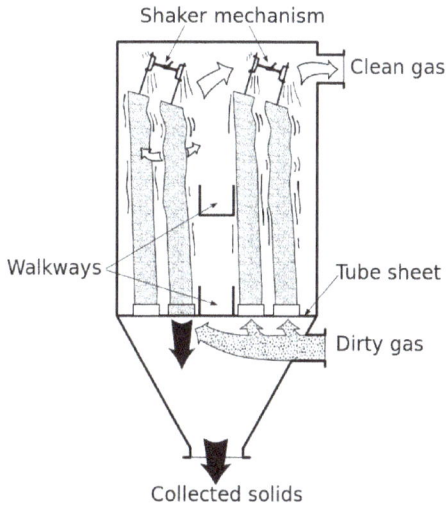

Mechanical Shaker Baghouse

Reverse Air Baghouse

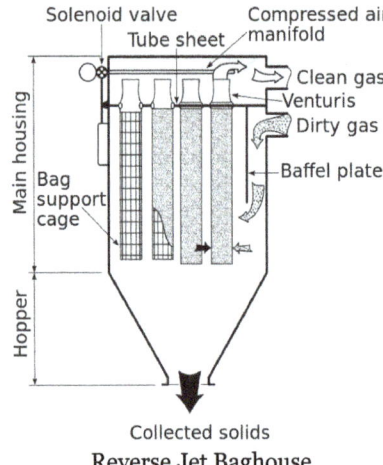

Reverse Jet Baghouse

Baghouses are classified by the cleaning method used. The three most common types of baghouses are mechanical shakers, reverse gas, and pulse jet.

Mechanical Shakers

In mechanical-shaker baghouses, tubular filter bags are fastened onto a cell plate at the bottom of the baghouse and suspended from horizontal beams at the top. Dirty gas enters the bottom of the baghouse and passes through the filter, and the dust collects on the inside surface of the bags.

Cleaning a mechanical-shaker baghouse is accomplished by shaking the top horizontal bar from which the bags are suspended. Vibration produced by a motor-driven shaft and cam creates waves in the bags to shake off the dust cake.

Shaker baghouses range in size from small, handshaker devices to large, compartmentalized units. They can operate intermittently or continuously. Intermittent units can be used when processes operate on a batch basis-when a batch is completed, the baghouse can be cleaned. Continuous processes use compartmentalized baghouses; when one compartment is being cleaned, the airflow can be diverted to other compartments.

In shaker baghouses, there must be no positive pressure inside the bags during the shake cycle. Pressures as low as 0.02 in. wg can interfere with cleaning.

The air to cloth ratio for shaker baghouses is relatively low, hence the space requirements are quite high. However, because of the simplicity of design, they are popular in the minerals processing industry.

Reverse air (R/A)

In reverse-air baghouses, the bags are fastened onto a cell plate at the bottom of the baghouse and suspended from an adjustable hanger frame at the top. Dirty gas flow normally enters the baghouse and passes through the bag from the inside, and the dust collects on the inside of the bags.

Reverse-air baghouses are compartmentalized to allow continuous operation. Before a cleaning cycle begins, filtration is stopped in the compartment to be cleaned. Bags are cleaned by injecting clean air into the dust collector in a reverse direction, which pressurizes the compartment. The pressure makes the bags collapse partially, causing the dust cake to crack and fall into the hopper below. At the end of the cleaning cycle, reverse airflow is discontinued, and the compartment is returned to the main stream.

The flow of the dirty gas helps maintain the shape of the bag. However, to prevent total collapse and fabric chafing during the cleaning cycle, rigid rings are sewn into the bags at intervals.

Space requirements for a reverse-air baghouse are comparable to those of a shaker baghouse; however, maintenance needs are somewhat greater.

Pulse Jet (Aka Reverse Jet)

In reverse-pulse-jet baghouses, individual bags are supported by a metal cage (filter cage), which is fastened onto a cell plate at the top of the baghouse. Dirty gas enters from the bottom of the baghouse and flows from outside to inside the bags. The metal cage prevents collapse of the bag.

Bags are cleaned by a short burst of compressed air injected through a common manifold over a row of bags. The compressed air is accelerated by a venturi nozzle mounted at the reverse-jet baghouse top of the bag. Since the duration of the compressed-air burst is short (0.1s), it acts as a rapidly moving air bubble, traveling through the entire length of the bag and causing the bag surfaces to flex. This flexing of the bags breaks the dust cake, and the dislodged dust falls into a storage hopper below.

Reverse-pulse-jet dust collectors can be operated continuously and cleaned without interruption of flow because the burst of compressed air is very small compared with the total volume of dusty air through the collector. Because of this continuous-cleaning feature, reverse-jet dust collectors are usually not compartmentalized.

The short cleaning cycle of reverse-jet collectors reduces recirculation and redeposit of dust. These collectors provide more complete cleaning and reconditioning of bags than shaker or reverse-air cleaning methods. Also, the continuous-cleaning feature allows them to operate at higher air-to-cloth ratios, so the space requirements are lower.

This cleaning system works with the help of digital sequential timer attached to the fabric filter. this timer indicates the solenoid valve to inject the air to the blow pipe.

Cleaning Method Comparison

Type	Advantages	Disadvantages
Shaker	Have high collection efficiency for respirable dust	Have low air-to-cloth ratio (1.5 to 2 ft/min)
	Can use strong woven bags, which can withstand intensified cleaning cycle to reduce residual dust buildup	Cannot be used in high temperatures
	Simple to operate	Require large amounts of space
	Have low pressure drop for equivalent collection efficiencies	Need large numbers of filter bags
		Consist of many moving parts and require frequent maintenance
		Personnel must enter baghouse to replace bags, creating potential for exposure to toxic dust
		Can result in reduced cleaning efficiency if even a slight positive pressure exists inside bags
Reverse air	Have high collection efficiency for respirable dust	Have low air-to-cloth ratio (1 to 2 ft/min)
	Are preferred for high temperatures due to gentle cleaning action	Require frequent cleaning because of gentle cleaning action
	Have low pressure drop for equivalent collection efficiencies	Have no effective way to remove residual dust buildup
		Cleaning air must be filtered
		Require personnel to enter baghouse to replace bags which creates potential for toxic dust exposure
Pulse jet (Reversed Jet)	Have high collection efficiency for respirable dust	Require use of dry compressed air
	Can have high air-to-cloth ratio (6 to 10 ft/min)	May not be used readily in high temperatures unless special fabrics are used
	Have increased efficiency and minimal residual dust buildup due to aggressive cleaning action	Cannot be used if high moisture content or humidity levels are present in the exhaust gases
	Can clean continuously	
	Can use strong woven bags	
	Have lower bag wear	
	Have small size and fewer bags because of high air-to-cloth ratio	
	Some designs allow bag changing without entering baghouse	
	Have low pressure drop for equivalent collection efficiencies	

Cleaning Considerations

Sonic Horns

Some baghouses have sonic horns installed to provide supplementary vibration cleaning energy. The horns, which generate high intensity, low frequency sounds waves, are turned on just before or at the start of the cleaning cycle to help break the bonds between particles on the filter media surface and aid in dust removal.

Cleaning Sequences

Two main sequence types are used to clean baghouses:

- Intermittent (periodic) cleaning

- Continuous cleaning.

Intermittently cleaned baghouses are composed of many compartments or sections. One at a time, each compartment is periodically closed off from the incoming dirty gas stream, cleaned, and then brought back online. While the individual compartment is out of place, the gas stream is diverted from the compartment's area. This makes shutting down the production process unnecessary during cleaning cycles.

Continuously cleaned baghouse compartments are always online for automatic filtering. A blast of compressed air momentarily interrupts the collection process to clean the bag. This is known as pulse jet cleaning. Pulse jet cleaning does not require taking compartments offline. Continuously cleaned baghouses are designed to prevent complete shutdown during bag maintenance and failures to the primary system.

Performance

Baghouse performance is contingent upon inlet and outlet gas temperature, pressure drop, opacity, and gas velocity. The chemical composition, moisture, acid dew point, and particle loading and size distribution of the gas stream are essential factors as well

- Gas Temperature - Fabrics are designed to operate within a certain range of temperature. Fluctuation outside of these limits even for a small period of time, can weaken, damage, or ruin the bags.

- Pressure Drop - Baghouses operate most effectively within a certain pressure drop range. This spectrum is based on a specific gas volumetric flow rate.

- Opacity - Opacity measures the quantity of light scattering that occurs as a result of the particles in a gas stream. Opacity is not an exact measurement of the concentration of particles; however, it is a good indicator of the amount of dust leaving the baghouse.

- Gas Volumetric Flow Rate - Baghouses are created to accommodate a range of gas flows. An increase in gas flow rates causes an increase in operating pressure drop and air-to-cloth ratio. These increases require the baghouse to work more strenuously, resulting in more frequent cleanings and high particle velocity, two factors that shorten bag life.

Design Variables

Pressure drop, filter drag, air-to-cloth ratio, and collection efficiency are essential factors in the design of a baghouse.

- Pressure drop is the resistance to air flow across the baghouse. A high pressure drop corresponds with a higher resistance to airflow. Pressure drop is calculated by determining the difference in total pressure at two points, typically the inlet and outlet.

- Filter drag is the resistance across the fabric-dust layer. It is the pressure drop per unit of velocity.

- An understanding of the term air-to-cloth ratio is vital to understand the mechanics of any baghouse system regardless of the exact type used. This ratio is defined as the amount of air or process gas entering the Baghouse divided by the sq. ft of cloth in the Baghouse. Units of measure are $(ft^3/min)/ft^2$ or $(cm^3/sec)/cm^2$.

- Commonly, baghouses are designed with 99.9% collection efficiency. Often, cleaned air is recirculated back into the plant for heating.

Filter Media

Fabric filter bags (sometimes referred to as envelopes) are oval or round tubes, typically 15–30 feet and 5 to 12 inches in diameter, made of woven or felted material. Depending on chemical and/or moisture content of the gas stream, its temperature, and other conditions, bags may be constructed out of cotton, nylon, polyester, fiberglass or other materials.

Nonwoven materials are either felted or membrane. Nonwoven materials are attached to a woven backing (scrim). Felted filters contain randomly placed fibers supported by a woven backing material (scrim). In a membrane filter, a thin, porous membrane is bound to the scrim. High energy cleaning techniques such as pulse jet require felted fabrics.

Woven filters have a definite repeated pattern. Low energy cleaning methods such as shaking or reverse air allow for woven filters. Various weaving patterns such as plain weave, twill weave, or sateen weave, increase or decrease the amount of space between individual fibers. The size of the space affects the strength and permeability of the fabric. A tighter weave corresponds with low permeability and, therefore, more efficient capture of fine particles.

Reverse air bags have anti-collapse rings sewn into them to prevent pancaking when cleaning energy is applied. Pulse jet filter bags are supported by a metal cage, which keeps the fabric taut. To lengthen the life of filter bags, a thin layer of PTFE (teflon) membrane may be adhered to the filtering side of the fabric, keeping dust particles from becoming embedded in the filter media fibers.

Some baghouses use pleated cartridge filters, similar to what is found in home air filtration systems.

Components

- Bags, fabric & support
- Housing or shell
- Collection hoppers
- Discharge devices
- Filter cleaning device
- Fan
- Compressor
- Timer Unit

References

- "NATA | National-Scale Air Toxics Assessments | Technology Transfer Network Air Technical Web Site | US EPA". Epa.gov. 2006-06-28. Retrieved 2012-12-11
- Provost, E; Madhloum, N; Int Panis, L; De Boever, P; Nawrot, TS (May 2015). "Carotid intima-media thickness, a marker of subclinical atherosclerosis, and particulate air pollution exposure: the meta-analytical evidence". PLoS ONE. 10 (5): e0127014. PMC 4430520. PMID 25970426. doi:10.1371/journal.pone.0127014
- Mateen, F. J.; Brook, R. D. (2011). "Air Pollution as an Emerging Global Risk Factor for Stroke". JAMA. 305 (12): 1240–1241. PMID 21427378. doi:10.1001/jama.2011.352
- Davis, Devra (2002). When Smoke Ran Like Water: Tales of Environmental Deception and the Battle Against Pollution. Basic Books. ISBN 0-465-01521-2
- European Court of Justice, CURIA (2008). "PRESS RELEASE No 58/08 Judgment of the Court of Justice in Case C-237/07" (PDF). Retrieved 24 January 2015
- Gehring, U.; Wijga, A. H.; Brauer, M.; Fischer, P.; de Jongste, J. C.; Kerkhof, M.; Brunekreef, B. (2010). "Traffic-related air pollution and the development of asthma and allergies during the first

8 years of life". American Journal of Respiratory and Critical Care Medicine. 181 (6): 596–603. doi:10.1164/rccm.200906-0858OC

- Committee on Environmental Health (2004). "Ambient Air Pollution: Health Hazards to Children". Pediatrics. 114 (6): 1699–1707. PMID 15574638. doi:10.1542/peds.2004-2166

- Noyes, Robert (1991). Handbook of Pollution Control Processes. Noyes Publications. ISBN 9780815512905. Retrieved 6 August 2013

- European Commission. "Air quality: Commission sends final warning to UK over levels of fine particle pollution". Archived from the original on 11 May 2011. Retrieved 7 April 2011

- Health; Occupational; Thoracic Society, American (1996). "[Comparative Study Review]". American Journal of Respiratory and Critical Care Medicine. 153 (1): 3–50

- Meselson M, Guillemin J, Hugh-Jones M, et al. (November 1994). "The Sverdlovsk anthrax outbreak of 1979" (PDF). Science. 266 (5188): 1202–8. PMID 7973702. doi:10.1126/science.7973702

- House of Commons Environmental Audit Committee (2010). "Environmental Audit Committee - Fifth Report Air Quality". Retrieved 24 January 2015

Permissions

All chapters in this book are published with permission under the Creative Commons Attribution Share Alike License or equivalent. Every chapter published in this book has been scrutinized by our experts. Their significance has been extensively debated. The topics covered herein carry significant information for a comprehensive understanding. They may even be implemented as practical applications or may be referred to as a beginning point for further studies.

We would like to thank the editorial team for lending their expertise to make the book truly unique. They have played a crucial role in the development of this book. Without their invaluable contributions this book wouldn't have been possible. They have made vital efforts to compile up to date information on the varied aspects of this subject to make this book a valuable addition to the collection of many professionals and students.

This book was conceptualized with the vision of imparting up-to-date and integrated information in this field. To ensure the same, a matchless editorial board was set up. Every individual on the board went through rigorous rounds of assessment to prove their worth. After which they invested a large part of their time researching and compiling the most relevant data for our readers.

The editorial board has been involved in producing this book since its inception. They have spent rigorous hours researching and exploring the diverse topics which have resulted in the successful publishing of this book. They have passed on their knowledge of decades through this book. To expedite this challenging task, the publisher supported the team at every step. A small team of assistant editors was also appointed to further simplify the editing procedure and attain best results for the readers.

Apart from the editorial board, the designing team has also invested a significant amount of their time in understanding the subject and creating the most relevant covers. They scrutinized every image to scout for the most suitable representation of the subject and create an appropriate cover for the book.

The publishing team has been an ardent support to the editorial, designing and production team. Their endless efforts to recruit the best for this project, has resulted in the accomplishment of this book. They are a veteran in the field of academics and their pool of knowledge is as vast as their experience in printing. Their expertise and guidance has proved useful at every step. Their uncompromising quality standards have made this book an exceptional effort. Their encouragement from time to time has been an inspiration for everyone.

The publisher and the editorial board hope that this book will prove to be a valuable piece of knowledge for students, practitioners and scholars across the globe.

Index

www.ingramcontent.com/pod-product-compliance
Lightning Source LLC
Chambersburg PA
CBHW080403190526
45161CB00003B/113